Lecture Notes of the Institute for Computer Sciences, Social Informatics and Telecommunications Engineering 212

More information about this series at http://www.springer.com/series/8197

Lingjie Duan · Anibal Sanjab
Husheng Li · Xu Chen
Donatello Materassi · Rachid Elazouzi (Eds.)

Game Theory for Networks

7th International EAI Conference, GameNets 2017
Knoxville, TN, USA, May 9, 2017
Proceedings

 Springer

Editors

Lingjie Duan
Engineering Systems and Design Pillar
Singapore University of Technology and
  Design
Dover
Singapore

Xu Chen
School of Data and Computer Science,
  Higher Education Mega Center
Sun Yat-sen University
Guangzhou
China

Anibal Sanjab
Virginia Tech
Blacksburg, VA
USA

Donatello Materassi
University of Tennessee
Knoxville, TN
USA

Husheng Li
Electrical Engineering/Computer Sciences
The University of Tennessee
Knoxville, TN
USA

Rachid Elazouzi
Laboratoire Informatique d'Avignon (LIA)
University of Avignon
Avignon
France

ISSN 1867-8211   ISSN 1867-822X   (electronic)
Lecture Notes of the Institute for Computer Sciences, Social Informatics
and Telecommunications Engineering
ISBN 978-3-319-67539-8   ISBN 978-3-319-67540-4   (eBook)
DOI 10.1007/978-3-319-67540-4

Library of Congress Control Number: 2017953401

Printed on acid-free paper

This Springer imprint is published by Springer Nature
The registered company is Springer International Publishing AG
The registered company address is: Gewerbestrasse 11, 6330 Cham, Switzerland

Preface

The 7th EAI International Conference on Game Theory for Networks (GameNets 2017) was held in the city of Knoxville, Tennessee, USA, on May 9, 2017. The conference attracted researchers and practitioners sharing a deep interest in the field of game theory for networks. Game theory has proven to provide indispensable tools enabling the analysis, modeling, and design of traditional as well as emerging complex networks. In this regard, the mission of GameNets 2017 was to introduce novel advancements in the research, development, and design of game-theoretic tools for networks and to draw future directions that this research must take to cope with the ever-growing complexity of modern networks.

The conference included 15 papers whose scope ranges from advancing fundamental game-theoretic concepts to focusing on various prominent network-based applications in the fields of the smart electric grid, the Internet of Things, social networks, network security, mobile service markets, and epidemic control. The conference included two keynote addresses by Prof. Vincent Poor and Prof. Eitan Altman, whose keynotes focused, respectively, on the use of game theory in smart grids and network neutrality analyses.

We would like to express our gratitude to all the authors for their submissions and contributions as well as to the Technical Program Committee and the reviewers who performed and supervised the review process.

We would also like to thank the European Alliance for Innovations (EAI), whose support was indispensable to the success of GameNets 2017.

August 2017

Lingjie Duan
Anibal Sanjab
Husheng Li
Xu Chen
Donatello Materassi
Rachid Elazouzi

Organization

Steering Committee

Steering Committee Chair

Imrich Chlamtac   Create-Net, Italy

Steering Committee Members

Athanasios Vasilakos   Kuwait University, Kuwait
Victor C.M. Leung   UBC, Canada
Haijun Zhang   UBC, Canada

Organizing Committee

General Chair

Husheng Li   The University of Tennessee, USA

General Co-chair

Xu Chen   University of Göttingen, Germany

Technical Program Committee Co-chairs

Donatello Materassi   The University of Tennessee, USA
Rachid Elazouzi   University of Avignon, France
Francesco De Pellegrini   Fondazione Bruno Kessler (FBK), Italy

Workshops Chair

Ju Bin Song   Kyung Hee University, South Korea

Publicity and Social Media Chair

Lin Gao   Harbin Institute of Technology, China

Sponsorship and Exhibits Chair

Chao Tian   The University of Tennessee, USA

Publications Chairs

Lingjie Duan   Singapore University of Technology and Design,
                                Singapore
Anibal Sanjab   Virginia Polytechnic Institute and State University,
                                USA

Local Chairs

Derek Sprintz   The University of Tennessee, USA
Yifan Wang   The University of Tennessee, USA

Web Chair

Liang Li   The University of Tennessee, USA

Conference Coordinator

Lenka Koczová   EAI (European Alliance for Innovation)

Technical Program Committee

Silva Allende Alonso	Nokia, Paris, France
Haddad Majed	University of Avignon, France
Hamidou Tembine	New York University, Abu Dhabi
Xu Yuedong	Fudan University, Shanghai, China
Mo Jeonghoon	Yonsei University, Seoul, South Korea
Daniel Sadoc	Federal University of Rio de Janeiro, Brazil
Avrachenkov Konstantin	Inria Sophia Antiplois, France
Coupechoux Marceau	Telecom ParisSud, Paris, France
Kesidis George	The Pennsylvania State University, University Park, USA
Yang Dejun	Colorado School of Mines, USA
Elias Jocelyne	Paris Descartes University, Paris, France
Koutsopoulos Iordanis	Athens University of Economics and Business, Athens, Greece
Al daoud Ashraf	University of Toronto, Canada
Essaid Sabir	ENSEM, Casablanca, Morocco
Saad Walid	Virginia Tech, Blacksburg, USA
Pavel Lacra	University of Toronto, Canada
Zhu Quanyan	NYU, New York, USA

Contents

Invited Papers

Games in Networks

Nash Equilibrium Seeking with Non-doubly Stochastic Communication Weight Matrix

Farzad Salehisadaghiani and Lacra Pavel[(✉)]

Department of Electrical and Computer Engineering, University of Toronto,
10 King's College Road, Toronto, ON M5S 3G4, Canada
farzad.salehisadaghiani@mail.utoronto.ca, pavel@control.utoronto.ca

Abstract. A distributed Nash equilibrium seeking algorithm is presented for networked games. We assume an incomplete information available to each player about the other players' actions. The players communicate over a strongly connected digraph to send/receive the estimates of the other players' actions to/from the other local players according to a gossip communication protocol. Due to asymmetric information exchange between the players, a non-doubly (row) stochastic weight matrix is defined. We show that, due to the non-doubly stochastic property, there is no exact convergence. Then, we present an almost sure convergence proof of the algorithm to a Nash equilibrium of the game. Moreover, we extend the algorithm for graphical games in which all players' cost functions are only dependent on the local neighboring players over an interference digraph. We design an assumption on the communication digraph such that the players are able to update all the estimates of the players who interfere with their cost functions. It is shown that the communication digraph needs to be a superset of a transitive reduction of the interference digraph. Finally, we verify the efficacy of the algorithm via a simulation on a social media behavioral case.

1 Introduction

The problem of finding a Nash equilibrium (NE) of a networked game has recently drawn many attentions. The players who participate in this game aim to minimize their own cost functions selfishly by making decision in response to other players' actions. Each player in the network has only access to local information of the neighbors. Due to the imperfect information available to players, they maintain an estimate of the other players' actions and communicate over a communication graph in order to exchange the estimates with local neighbors. Using this information, players update their actions and estimates.

In many algorithms in the context of NE seeking problems, it is assumed that the communications between the players are symmetric in the sense that the players who are in communication can exchange their information altogether and update their estimates at the same time. This, in general, leads to a doubly

This work was supported by an NSERC Discovery Grant.

© ICST Institute for Computer Sciences, Social Informatics and Telecommunications Engineering 2017
L. Duan et al. (Eds.): GameNets 2017, LNICST 212, pp. 3–15, 2017.
DOI: 10.1007/978-3-319-67540-4_1

stochastic communication weight matrix which preserves the global average of the estimates over time. However, there are many real-world applications in which symmetric communication is not of interest or may be an undesired feature in applications such as sensor network.

Literature review. Our work is related to the literature on Nash games and distributed NE seeking algorithms [1,4,11,16,17]. A distributed algorithm is proposed in [18] to compute a generalized NE of the game for a complete communication graph. In [7], an algorithm is provided to find an NE of *aggregative games* for a partial communication graph but complete interference graph. This algorithm is extended by [12] for a more general class of games in which the players' cost functions does not necessarily depend on the aggregate of players' actions. It is further generalized for the partial interference graph in [13]. For a *two-network zero-sum game* [5] considers a distributed algorithm for NE seeking. To find distributed algorithms for games with local-agent utility functions, a methodology is presented in [8] based on state-based *potential games*.

Gossip-based communication has been widely used in synchronous and asynchronous algorithms in consensus and distributed optimization problems [2,3,9]. In [9], a gossip algorithm is designed for a distributed broadcast-based optimization problem. An almost-sure convergence is provided due to the non-doubly stochasticity of the communication matrix. In [2], a broadcast gossip algorithm is studied to compute the average of the initial measurements which is proved to converge almost surely to a consensus.

Contributions. We propose an asynchronous gossip-based algorithm to find an NE of a distributed game over a communication digraph. We assume that players send/receive information to/from their local out/in-neighbors over a strongly connected communication digraph. Players update their own actions and estimates based on the received information. We prove an almost sure convergence of the algorithm to the NE of the game. *Unlike in the undirected case* [12,13], *herein we cannot exploit the doubly stochastic property for the communication weight matrix due to asymmetric information exchange. Non-doubly stochastic property leads to have total average of the players' estimates not preserved over time. This was one of the critical steps in the convergence proof in* [12,13].

Moreover, we extend the algorithm for graphical games in which the players' cost functions may be interfered by any subset of players' (not necessarily all the players') actions. The locality of cost functions is specified by an interference digraph which marks the pair of players who interfere one with another. In order to have a convergent algorithm, we design an assumption on the communication digraph by which there exists a lower bound on the communication digraph which is a transitive reduction of the interference digraph. By this assumption, it is proved that all the players are able to exchange and update all the estimates of the actions interfering with their cost functions.

The proofs are omitted due to space limitations, and are available in [14].

2 Problem Statement: Game with a Complete Interference Digraph

Consider a multi-player game in a network with a set of players V. The interference of players' actions on the cost functions is represented by a complete *interference digraph* $G(V, E)$, with E marking the pair of players that interfere one with another. Note that for a complete digraph every pair of distinct nodes is connected by a pair of unique edges (one in each direction).

The game is denoted by $\mathcal{G}(V, \Omega_i, J_i)$ and defined over

- $V = \{1, \ldots, N\}$: Set of players,
- $\Omega_i \subset \mathbb{R}$: Action set of player i, $\forall i \in V$ with $\Omega = \prod_{i \in V} \Omega_i \subset \mathbb{R}^N$ the action set of all players,
- $J_i : \Omega \to \mathbb{R}$: Cost function of player i, $\forall i \in V$,

In the following we define a few notations for players' actions.

- $x = (x_i, x_{-i}) \in \Omega$: All players actions,
- $x_i \in \Omega_i$: Player i's action, $\forall i \in V$ and $x_{-i} \in \Omega_{-i} := \prod_{j \in V \setminus \{i\}} \Omega_j$: All other players' actions except i.

The game is defined as a set of N simultaneous optimization problems as follows:

$$\begin{cases} \underset{y_i}{\text{minimize}} & J_i(y_i, x_{-i}) \\ \text{subject to} & y_i \in \Omega_i \end{cases} \quad \forall i \in V. \tag{1}$$

Each problem is run by an individual player and its solution is dependent on the solution of the other problems. The objective is to find an NE of this game which is defined as follows:

Definition 1. *Consider an N-player game $\mathcal{G}(V, \Omega_i, J_i)$, each player i minimizing the cost function $J_i : \Omega \to \mathbb{R}$. A vector $x^* = (x_i^*, x_{-i}^*) \in \Omega$ is called an NE of this game if*

$$J_i(x_i^*, x_{-i}^*) \leq J_i(x_i, x_{-i}^*) \quad \forall x_i \in \Omega_i, \ \forall i \in V. \tag{2}$$

We state a few assumptions for the existence and the uniqueness of an NE.

Assumption 1. *For every $i \in V$,*

- *Ω_i is non-empty, compact and convex,*
- *$J_i(x_i, x_{-i})$ is C^1 in x_i, continuous in x and convex in x_i for every x_{-i}.*

The compactness of Ω implies that $\forall i \in V$ and $x \in \Omega$,

$$\|\nabla_{x_i} J_i(x)\| \leq C, \quad \text{for some } C > 0. \tag{3}$$

Let $F : \Omega \to \mathbb{R}^N$, $F(x) := [\nabla_{x_i} J_i(x)]_{i \in V}$ be the pseudo-gradient vector of the cost functions (game map).

Assumption 2. *F is strictly monotone, $(F(x) - F(y))^T (x - y) > 0 \quad \forall x, y \in \Omega, \ x \neq y$.*

Assumption 3. *$\nabla_{x_i} J_i(x_i, u)$ is Lipschitz continuous in x_i, for every fixed $u \in \Omega_{-i}$ and for every $i \in V$, i.e., there exists $\sigma_i > 0$ such that*

$$\|\nabla_{x_i} J_i(x_i, u) - \nabla_{x_i} J_i(y_i, u)\| \leq \sigma_i \|x_i - y_i\| \quad \forall x_i, y_i \in \Omega_i.$$

Moreover, $\nabla_{x_i} J_i(x_i, u)$ is Lipschitz continuous in u with a Lipschitz constant $L_i > 0$ for every fixed $x_i \in \Omega_i, \forall i \in V$.

In game (1), the only information available to each player i is J_i and Ω. Thus, each player maintains an estimate of the other players actions and exchanges those estimates with the neighbors to update them. A *communication digraph* $G_C(V, E_C)$ is defined where $E_C \subseteq V \times V$ denotes the set of communication links between the players. $(i, j) \in E_C$ if and only if player i sends his information to player j. Note that $(i, j) \in E_C$ does not necessarily imply $(j, i) \in E_C$. The set of in-neighbors of player i in G_C, denoted by $N_C^{\text{in}}(i)$, is defined as $N_C^{\text{in}}(i) := \{j \in V | (j, i) \in E_C\}$. The following assumption on G_C is used.

Assumption 4. *G_C is strongly connected.*

Our objective is to find an algorithm for computing an NE of $\mathcal{G}(V, \Omega_i, J_i)$ using only imperfect information over the communication digraph $G_C(V, E_C)$.

3   Asynchronous Gossip-Based Algorithm

We propose a projected gradient-based algorithm using an asynchronous gossip-based method as in [12]. The algorithm is inspired by [12] except that the communications are supposed to be directed in a sense that the information exchange is considered over a directed path. Our challenge here is to deal with the asymmetric communications between the players. This makes the convergence proof dependent on a *non-doubly stochastic weight matrix*, whose properties need to be investigated and proved. The algorithm is elaborated as follows:

1- **Initialization Step:** Each player i maintains an initial *temporary* estimate $\tilde{x}^i(0) \in \Omega$ for all players. Let $\tilde{x}_j^i(0) \in \Omega_j \subset \mathbb{R}$ be player i's initial temporary estimate of player j's action, for $i, j \in V$.

2- **Gossiping Step:** At iteration k, player i_k becomes active uniformly at random and selects a communication in-neighbor indexed by $j_k \in N_C^{\text{in}}(i_k)$ uniformly at random. Let $\tilde{x}^i(k) \in \Omega \subset \mathbb{R}^N$ be player i's temporary estimate at iteration k. Then player j_k sends his temporary estimate $\tilde{x}^{j_k}(k)$ to player i_k. After receiving the information, player i_k constructs his final estimate of all players. Let $\hat{x}_j^i(k) \in \Omega_j \subset \mathbb{R}$ be player i's final estimate of player j's action, for $i, j \in V$. The final estimates are computed as in the following:

1. Players i_k's final estimate:

$$\begin{cases} \hat{x}_{i_k}^{i_k}(k) = \tilde{x}_{i_k}^{i_k}(k) \\ \hat{x}_{-i_k}^{i_k}(k) = \frac{\tilde{x}_{-i_k}^{i_k}(k) + \tilde{x}_{-i_k}^{j_k}(k)}{2}. \end{cases} \tag{4}$$

Note that $\tilde{x}_i^i(k) = x_i(k)$ for all $i \in V$.

2. For all other players $i \neq i_k$, the temporary estimate is maintained, i.e.,

$$\hat{x}^i(k) = \tilde{x}^i(k), \quad \forall i \neq i_k. \tag{5}$$

We use communication weight matrix $W(k) := [w_{ij}(k)]_{i,j \in V}$ to obtain a compact form of the gossip protocol. $W(k)$ is a *non-doubly (row) stochastic weight matrix* defined as follows:

$$W(k) = I_N - \frac{e_{i_k}(e_{i_k} - e_{j_k})^T}{2}, \tag{6}$$

where $e_i \in \mathbb{R}^N$ is a unit vector. Note that $W(k)$ is different from the doubly stochastic one used in [12]. The non-doubly (row) stochasticity of $W(k)$ is translated into:

$$W(k)\mathbf{1}_N = \mathbf{1}_N, \quad \mathbf{1}_N^T W(k) \neq \mathbf{1}_N^T. \tag{7}$$

Let $\bar{x}(k) = [\bar{x}^1(k), \ldots, \bar{x}^N(k)]^T \in \Omega^N$ be an intermediary variable such that

$$\bar{x}(k) = (W(k) \otimes I_N)\tilde{x}(k), \tag{8}$$

where $\tilde{x}(k) = [\tilde{x}^1(k), \ldots, \tilde{x}^N(k)]^T \in \Omega^N$ is the overall temporary estimate at k. Using (6) one can combine (4) and (5) in a compact form of $\hat{x}_{-i_k}^{i_k}(k) = \bar{x}_{-i_k}^{i_k}(k)$ and $\hat{x}^i(k) = \bar{x}^i(k)$ for $\forall i \neq i_k$.

3- **Local Step:** At this moment all the players update their actions according to a projected gradient-based method. Let $\bar{x}^i = (\bar{x}_i^i, \bar{x}_{-i}^i) \in \Omega, \forall i \in V$ with $\bar{x}_i^i \in \Omega_i$ be the intermediary variable associated to player i. Because of imperfect information available to player i, he uses $\bar{x}_{-i}^i(k)$ and updates his action as follows: if $i = i_k$,

$$x_i(k+1) = T_{\Omega_i}[x_i(k) - \alpha_{k,i} \nabla_{x_i} J_i(x_i(k), \bar{x}_{-i}^i(k))], \tag{9}$$

otherwise, $x_i(k+1) = x_i(k)$. In (9), $T_{\Omega_i} : \mathbb{R} \to \Omega_i$ is an Euclidean projection and $\alpha_{k,i}$ are diminishing step sizes such that $\sum_{k=1}^{\infty} \alpha_{k,i}^2 < \infty, \sum_{k=1}^{\infty} \alpha_{k,i} = \infty \, \forall i \in V$. The players use their updated actions to update their temporary estimates as follows:

$$\tilde{x}^i(k+1) = \bar{x}^i(k) + (x_i(k+1) - \bar{x}_i^i(k))e_i, \quad \forall i \in V. \tag{10}$$

At this point, the players are ready to begin a new iteration from step 2. We elaborate on the non-doubly stochasticity of $W(k)$ from two perspectives.

1. **Design:** By the row (non-doubly) stochastic property of $W(k)$, the temporary estimates remain at consensus subspace once they reach there. This can be verified by (8) when $\tilde{x}(k) = \mathbf{1}_N \otimes \boldsymbol{\alpha}$ for an $N \times 1$ vector $\boldsymbol{\alpha}$, since,

$$\bar{x}(k) = (W(k) \otimes I_N)(\mathbf{1}_N \otimes \boldsymbol{\alpha}) = \mathbf{1}_N \otimes \boldsymbol{\alpha}. \tag{11}$$

Equations (9), (10) and (11) imply that the consensus is maintained. On the other hand $W(k)$ *is not column-stochastic* which is a critical property used in [12]. This implies that the average of temporary estimates is not equal to the average of $\bar{x}$. Indeed by (8),

$$\frac{1}{N}(\mathbf{1}_N^T \otimes I_N)\bar{x}(k) = \frac{1}{N}(\mathbf{1}_N^T \otimes I_N)(W(k) \otimes I_N)\tilde{x}(k) \neq \frac{1}{N}(\mathbf{1}_N^T \otimes I_N)\tilde{x}(k). \tag{12}$$

Equations (9), (10) and (12) imply that the average of temporary estimates is not preserved for the next iteration. Thus, it is infeasible to obtain an exact convergence to the average consensus [2]. Instead, we show an almost sure (a.s.) convergence of the temporary estimates to an average consensus[1].

2. **Convergence Proof:** $\lambda_{\max}(W(k)^T W(k))$ is a key parameter in the proof (as in [9,12]). Unlike [12], the non-doubly stochastic property of $W(k)^T W(k)$ ends up in having $\lambda_{\max}(W(k)^T W(k)) > 1$. We resolve this issue in Lemma 1.

4   Convergence for Diminishing Step Sizes

In this section we prove convergence of the algorithm for diminishing step sizes. Consider a memory in which the history of the decision making is recorded. Let $\mathcal{M}_k$ denote the *sigma-field* generated by the history up to time $k - 1$ with

$$\mathcal{M}_0 = \{\tilde{x}^i(0), \ i \in V\}. \ \mathcal{M}_k = \mathcal{M}_0 \cup \left\{(i_l, j_l); 1 \leq l \leq k - 1\right\}, \quad \forall k \geq 2. \tag{13}$$

As explained in the design challenge in Sect. 3, we consider a.s. convergence. Convergence is shown in two parts. First, we prove a.s. convergence of the temporary estimate vectors $\tilde{x}^i$, to an average consensus, proved to be the vectors' average. Then we prove a.s. convergence of players' actions toward an NE.

Let $\tilde{x}(k)$ be the overall temporary estimate vector. The average of all temporary estimates at $T(k)$ is defined as follows:

$$Z(k) = \frac{1}{N}(\mathbf{1}_N^T \otimes I_N)\tilde{x}(k). \tag{14}$$

As mentioned in Sect. 3, the major difference between the proposed algorithm and the one in [12] is in using a non-doubly stochastic weight matrix $W(k)$ which was a key step. The following lemma is used to overcome these challenges.

---

[1] The same objective is followed by [9] to find a broadcast gossip algorithm (with non-doubly stochastic weight matrix) in the area of distributed optimization. However, in the proof of Lemma 2 ([9] page 1348) which is mainly dedicated to this discussion, the doubly stochasticity of $W(k)$ is used right after Eq. (22) which violates the main assumption on $W(k)$.

Lemma 1. *Let $Q(k) = (W(k) - \frac{1}{N}\mathbf{1}_N\mathbf{1}_N^T W(k))\otimes I_N$ and $W(k)$ be a non-doubly (row) stochastic weight matrix defined in (6) which satisfies (7). Let also $\gamma = \lambda_{\max}(\mathbb{E}[Q(k)^T Q(k)])$. Then $\gamma < 1$.*

Proof. See [14].

Theorem 1. *Let $\tilde{x}(k)$ be the stack vector with all temporary estimates of the players and $Z(k)$ be its average as in (14). Let also $\alpha_{k,max} = \max_{i\in V}\alpha_{k,i}$. Then under Assumptions 1, 4, the following hold.*

(i) $\sum_{k=0}^{\infty}\alpha_{k,max}\|\tilde{x}(k) - (\mathbf{1}_N \otimes I_N)Z(k)\| < \infty$ a.s.,
(ii) $\sum_{k=0}^{\infty}\|\tilde{x}(k) - (\mathbf{1}_N \otimes I_N)Z(k)\|^2 < \infty$ a.s.

Proof. The proof follows as in the proof of Theorem 1 in [12], but the critical step here is in using Lemma 1.

Corollary 1. *For the players' actions $x(k)$ and $\bar{x}(k)$, the following terms hold a.s. under Assumptions 1–4.*

(i) $\sum_{k=0}^{\infty}\alpha_{k,max}\|x(k) - Z(k)\| < \infty$ a.s.,   (ii) $\sum_{k=0}^{\infty}\|x(k) - Z(k)\|^2 < \infty$ a.s.,
(iii) $\sum_{k=0}^{\infty}\mathbb{E}\left[\|\bar{x}(k) - (\mathbf{1}_N \otimes I_N)Z(k)\|^2\Big|\mathcal{M}_k\right] < \infty$ a.s.

Proof. See [14].

Theorem 2. *Let $x(k)$ and x^* be the players' actions and the NE of $\mathcal{G}$, respectively. Under Assumptions 3–4, the sequence $\{x(k)\}$ generated by the algorithm converges to x^*, almost surely.*

Proof. The proof is similar to the proof of Theorem 2 in [12] based on Theorem 1. Theorem 2 verifies that the actions of the players converge a.s. toward the NE using the fact that the actions converge to a consensus subspace (Corollary 1).

5   Game with a Partial Interference Digraph

We extend the game defined in Sect. 2 to the case with partially coupled cost functions, such that cost functions may be interfered by the actions of any subset of players. The game is denoted by $\mathcal{G}(V, G_I, \Omega_i, J_i)$ where $G_I(V, E_I)$ is an interference digraph with E_I marking the players whose actions interfere with the other players' cost functions. We denote by $N_I^{in}(i) := \{j \in V|(j,i) \in E_I\}$, the set of in-neighbors of player i in G_I whose actions affect J_i and $\tilde{N}_I^{in}(i) := N_I^{in}(i) \cup \{i\}$.

The following assumption is considered for G_I.

Assumption 5. *G_I is strongly connected.*

The cost function of player i, J_i, $\forall i \in V$, is defined over $\Omega^i \to \mathbb{R}$ where $\Omega^i = \prod_{j\in\tilde{N}_I^{in}(i)} \Omega_j \subset \mathbb{R}^{|\tilde{N}_I^{in}(i)|}$ is the action set of players interfering with the cost function of player i. A few notations for players' actions are given:

- $x^i = (x_i, x^i_{-i}) \in \Omega^i$: All players' actions which interfere with J_i,
- $x^i_{-i} \in \Omega^i_{-i} := \prod_{j \in N^{in}_I(i)} \Omega_j$: Other players' actions which interfere with J_i.

Given x^i_{-i}, each player i aims to minimize his own cost function selfishly,

$$\begin{cases} \underset{y_i}{\text{minimize}} & J_i(y_i, x^i_{-i}) \\ \text{subject to} & y_i \in \Omega_i \end{cases} \quad \forall i \in V. \tag{15}$$

Known parameters to player i are as follows: (1) Cost function of player i, J_i and (2) Action set Ω^i. Note that this game setup is similar to the one in [13] except for a directed G_C used for asymmetric communications. Our first objective is to design an assumption on G_C such that all required information is communicated by the players after sufficiently many iterations. In other words, we ensure that player i, $\forall i \in V$ receives information from all the players whose actions interfere with his cost function.

Definition 2. *Transitive reduction: A digraph H is a transitive reduction of G which is obtained as follows: for all three vertices i, j, l in G such that edges (i, j), (j, l) are in G, (i, l) is removed from G.*

Note that transitive reduction is different from *maximal triangle-free spanning subgraph* which is used in Assumption 2 in [13].

Assumption 6. *The following holds for the communication graph G_C:*

- $G_{TR} \subseteq G_C \subseteq G_I$, where G_{TR} is a transitive reduction of G_I.

Lemma 2. *Let G_I and G_C satisfying Assumptions 5, 6. Then, $\forall i \in V$,*

$$\bigcup_{j \in N^{in}_C(i)} \left(N^{in}_I(i) \cap \tilde{N}^{in}_I(j) \right) = N^{in}_I(i). \tag{16}$$

Proof. See [14].

Remark 1. (16) verifies that using Assumptions 5, 6 the first objective is satisfied.

The assumptions for existence and uniqueness of an NE are Assumptions 1–3 with the cost functions adapted to G_I. Our second objective is to find an algorithm for computing an NE of $\mathcal{G}(V, G_I, \Omega_i, J_i)$ over $G_C(V, E_C)$ with partially coupled cost functions as described by the directed graph $G_I(V, E_I)$.

6   Asynchronous Gossip-Based Algorithm Adapted to G_I

The structure of the algorithm is similar to the one in Sect. 3. The steps are elaborated in the following:

1- **Initialization Step:**
  – $\tilde{x}^i(0) \in \Omega^i$: Player i's initial temporary estimate.
2- **Gossiping Step:**
  – $\tilde{x}^i_j(k) \in \Omega_j \subset \mathbb{R}$: Player i's temporary estimate of player j's action at k.
  – $\hat{x}^i_j(k) \in \Omega_j \subset \mathbb{R}$: Player i's final estimate of player j's action at k, for $i \in V$, $j \in \tilde{N}^{in}_I(i)$.
  – Final estimate construction:

$$\hat{x}^{i_k}_l(k) = \begin{cases} \frac{\tilde{x}^{i_k}_l(k) + \tilde{x}^{j_k}_l(k)}{2}, & l \in (N^{in}_I(i_k) \cap \tilde{N}^{in}_I(j_k)) \\ \tilde{x}^{i_k}_l(k), & l \in \tilde{N}^{in}_I(i_k) \backslash (N^{in}_I(i_k) \cap \tilde{N}^{in}_I(j_k)). \end{cases} \tag{17}$$

For

$$i \neq i_k,\ j \in \tilde{N}^{in}_I(i),\ \hat{x}^i_j(k) = \tilde{x}^i_j(k). \tag{18}$$

We suggest a compact form for gossip protocol by using $W^I(k)$.
Let for player i,

$$W^I(k) := I_m - \sum_{l \in (\tilde{N}^{in}_I(i_k) \cap \tilde{N}^{in}_I(j_k))} \frac{e_{s_{i_k l}}(e_{s_{i_k l}} - e_{s_{j_k l}})^T}{2}, \tag{19}$$

where $e_i \in \mathbb{R}^m$ is a unit vector. Note that $W^I(k)$ is different from the doubly stochastic one used in [13]. See [14] for the design of s_{ij} which is an index of the estimate vector element associated with player i's estimate of player j's action.
  – $\tilde{x}(k) := \left[\tilde{x}^{1^T}, \dots, \tilde{x}^{N^T}\right]^T$: Stack vector of all temporary estimates,
  – $\bar{x}(k) := W^I(k)\tilde{x}(k)$: Intermediary variable.
Using the intermediary variable, one can construct the final estimates as follows:

$$\hat{x}^i_{-i}(k) = [\bar{x}_{s_{ij}}(k)]_{j \in N^{in}_I(i)}. \tag{20}$$

3- **Local Step:** Player i updates his action as follows: If $i = i_k$, $x_i(k+1) = T_{\Omega_i}\left[x_i(k) - \alpha_{k,i} \nabla_{x_i} J_i(x_i(k), [\bar{x}_{s_{ij}}(k)]_{j \in N^{in}_I(i)})\right]$, otherwise,

$$x_i(k+1) = x_i(k), \tag{21}$$

Then he updates his temporary estimates:

$$\tilde{x}^i_j(k+1) = \begin{cases} \bar{x}_{s_{ij}}(k), & \text{if } j \neq i \\ x_i(k+1), & \text{if } j = i. \end{cases} \tag{22}$$

At this point, the players are ready to begin a new iteration from step 2.

7   Convergence of the Algorithm Adapted to G_I

Similar to Sect. 4, the convergence proof is split into two steps:

1. First, we prove a.s. convergence of $\tilde{x}(k) \subset \mathbb{R}^m$ to an average consensus which is shown to be the augmented average of all temporary estimate vectors. Let
   - $m_i^{\text{out}} := \deg_{G_I}^{\text{out}}(i) + 1$, where $\deg_{G_I}^{\text{out}}(i)$ is the out-degree of vertex i in G_I,
   - $1./\mathbf{m}^{\text{out}} := [\frac{1}{m_1^{\text{out}}}, \ldots, \frac{1}{m_N^{\text{out}}}]^T$,
   -

$$H := [\sum_{i:1 \in N_I^{\text{in}}(i)} e_{s_{i1}}, \ldots, \sum_{i:N \in N_I^{\text{in}}(i)} e_{s_{iN}}] \in \mathbb{R}^{m \times N}, \qquad (23)$$

where $i : j \in N_I^{\text{in}}(i)$ is all i's such that $j \in N_I^{\text{in}}(i)$. The *augmented average* of all temporary estimates is denoted by $Z^I(k) \in \mathbb{R}^m$ and defined as follows:

$$Z^I(k) := H \operatorname{diag}(1./\mathbf{m}^{\text{out}}) H^T \tilde{x}(k) \in \mathbb{R}^m. \qquad (24)$$

2. Secondly, we prove almost sure convergence of the players actions to an NE.

The proof depends on some key properties of W^I and H given in Lemmas 3, 4.

Lemma 3. *Let $W^I(k)$ and H be defined in (19) and (23). Then, $W^I(k)H = H$. This can be interpreted as the generalized row stochastic property of $W^I(k)$.*

Proof. See [14].

Lemma 4. *Let $Q^I(k) := W^I(k) - H \operatorname{diag}(1./\mathbf{m}^{out}) H^T W^I(k)$ and $\gamma^I = \lambda_{\max}\big(\mathbb{E}[Q^I(k)^T Q^I(k)]\big)$. Then $\gamma^I < 1$.*

Proof. See [14].

Theorem 3. *Let $\tilde{x}(k)$ be the stack vector with all temporary estimates of the players and $Z^I(k)$ be its average as in (24). Let also $\alpha_{k,max} = \max_{i \in V} \alpha_{k,i}$. Then under Assumptions 1', 5, 6, the following hold.*
(i) $\sum_{k=0}^{\infty} \alpha_{k,max} \|\tilde{x}(k) - Z^I(k)\| < \infty$ a.s.,   (ii) $\sum_{k=0}^{\infty} \|\tilde{x}(k) - Z^I(k)\|^2 < \infty$ a.s.

Proof. The proof uses Lemmas 3, 4 and is similar to the proof Theorem 1 in [15].

Corollary 2. *Let $z^I(k) := \operatorname{diag}(1./\mathbf{m}^{out}) H^T \tilde{x}(k) \in \mathbb{R}^N$ be the average of all players' temporary estimates. Under Assumptions 1', 5, 6 the following hold for players' actions $x(k)$ and $\bar{x}(k)$:*
(i) $\sum_{k=0}^{\infty} \alpha_{k,max} \|x(k) - z^I(k)\| < \infty$ a.s.,   (ii) $\sum_{k=0}^{\infty} \|x(k) - z^I(k)\|^2 < \infty$ a.s.,
(iii) $\sum_{k=0}^{\infty} \mathbb{E}\Big[\|\bar{x}(k) - Z^I(k)\|^2 \Big| \mathcal{M}_k\Big] < \infty$ a.s.

Proof. See [14].

Theorem 4. *Let $x(k)$ and x^* be all players' actions and the NE of $\mathcal{G}$, respectively. Under Assumptions 1'–3', 5, 6, the sequence $\{x(k)\}$ generated by the algorithm converges to x^*, almost surely.*

Proof. The proof uses Theorem 3 and is similar to the proof of Theorem 2 in [15].

8   Simulation Results

8.1   Social Media Behavior

In this example we aim to investigate social networking media for users' behavior. In such media like Facebook, Twitter and Instagram users are allowed to follow (or be friend with) the other users and post statuses, photos and videos or also share links and events. Depending on the type of social media, the way of communication is defined. For instance, in Instagram, friendship is defined unidirectional in a sense that either side could be only a follower and/or being followed. Recently, researchers at Microsoft have been studying the behavioral attitude of the users of Facebook as a giant and global network [10]. This study can be useful in many areas e.g. business (posting advertisements) and politics (posting for the purpose of presidential election campaign). Generating new status usually comes with the cost for the users such that if there is no benefit in posting status, the users don't bother to generate new ones. In any social media drawing others' attention is one of the most important motivation/stimulation to post status [6]. Our objective is to find the optimal rate of posting status for each user to draw more attention in his network. In the following, we make an information/attention model of a generic social media [6] and define a communication between users (G_C) and an interference graph between them (G_I).

Consider a social media network of N users. Each user i produces x_i unit of information that the followers can see in their news feeds. The users' communication network is defined by a strongly connected digraph G_C in which $(i) \rightarrow (j)$ means j is a follower of i or j receives x_i in his news feed. We also assume a strongly connected interference digraph G_I to show the influence of the users on the others. We assume that each user i's cost function is not only affected by the users he follows, but also by the users that his followers follow. The cost function of user i is denoted by J_i and consists of three parts: (1) $C_i(x_i) := h_i x_i$, $h_i > 0$ which is a cost that user i pays to produce x_i unit of information. (2) $f_i^1(x) := L_i \sqrt{\sum_{j \in N_C^{in}(i)} q_{ji} x_j}$, $L_i > 0$ which is a differentiable, increasing and concave utility function of user i from receiving information from his news feed with $f_i^1(0) = 0$ and q_{ji} represents follower i's interest in user j's information and L_i is a user-specific parameter. (3) $f_i^2(x) := \sum_{l:i \in N_C^{in}(l)} L_l \left(\sqrt{\sum_{j \in N_C^{in}(l)} q_{jl} x_j} - \sqrt{\sum_{j \in N_C^{in}(l) \setminus \{i\}} q_{jl} x_j} \right)$ which is an incremental utility function that each user obtains from receiving attention in his network with $f_i^2(x)|_{x_i=0} = 0$. Specifically, this function targets the amount of attention that each follower pays to the information of other users in his news feed. The total cost function for user i is then $J_i(x) = C_i(x_i) - f_i^1(x) - f_i^2(x)$. For this example, we consider 5 users in the social media whose network of followers G_C is given in Fig. 1(a). From G_C and taking J_i into account, one can construct G_I (Fig. 1(b)) in a way that the interferences among users are specified. Note that this is a reverse process of the one discussed in Sect. 5 because G_C is given as the network of followers and G_I is constructed from G_C. For the particular networks in Fig. 1(a, b), Assumptions 5, 6 hold. We then employ the algorithm

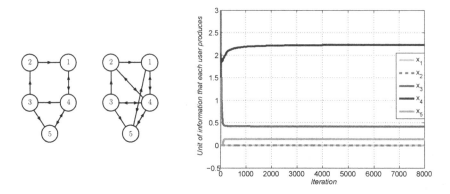

Fig. 1. (a) G_C (b) G_I (c) Convergence of the unit of information that each user produces to a NE over G_C.

in Sect. 6 to find an NE of this game for $h_i = 2$ and $L_i = 1.5 \ \forall i \in V$, and $q_{41} = q_{45} = 1.75$, $q_{32} = q_{43} = 2$ and the rest of $q_{ij} = 1$. The result is shown in Fig. 1(c). To analyze the NE $x^* = [0, 0, 0.42, 2.24, 0.14]^T$, one can realize from G_C that user 4 has 3 followers (users 1, 3 and 5), user 3 has 2 followers (users 2 and 5) and the rest has only 1 follower. Then, it is straightforward to predict that users 4 and 3 could draw more attentions and produce more information.

References

1. Alpcan, T., Başar, T.: Distributed algorithms for Nash equilibria of flow control games. In: Nowak, A.S., Szajowski, K. (eds.) Advances in Dynamic Games, pp. 473–498. Springer, Boston (2005). doi:10.1007/0-8176-4429-6_26
2. Aysal, T.C., Yildiz, M.E., Sarwate, A.D., Scaglione, A.: Broadcast gossip algorithms for consensus. Trans. Sig. Process. **57**(7), 2748–2761 (2009)
3. Fagnani, F., Zampieri, S.: Randomized consensus algorithms over large scale networks. IEEE J. Sel. Areas Commun. **26**(4), 634–649 (2008)
4. Frihauf, P., Krstic, M., Basar, T.: Nash equilibrium seeking in noncooperative games. IEEE Trans. Autom. Control **57**(5), 1192–1207 (2012)
5. Gharesifard, B., Cortes, J.: Distributed convergence to Nash equilibria in two-network zero-sum games. Automatica **49**(6), 1683–1692 (2013)
6. Goel, A., Ronaghi, F.: A game-theoretic model of attention in social networks. In: Bonato, A., Janssen, J. (eds.) WAW 2012. LNCS, vol. 7323, pp. 78–92. Springer, Heidelberg (2012). doi:10.1007/978-3-642-30541-2_7
7. Koshal, J., Nedic, A., Shanbhag, U.V.: A gossip algorithm for aggregative games on graphs. In: IEEE 51st Conference on Decision and Control, pp. 4840–4845 (2012)
8. Li, N., Marden, J.R.: Designing games for distributed optimization. IEEE J. Sel. Top. Sig. Process. **7**(2), 230–242 (2013)
9. Nedic, A.: Asynchronous broadcast-based convex optimization over a network. IEEE Trans. Autom. Control **56**(6), 1337–1351 (2011)
10. Olson, P.: Microsoft uses facebook as giant 'lab' to study game theory. Forbes (2011). http://www.forbes.com/sites/parmyolson/2011/10/12/microsoft-uses-facebook-as-giant-lab-to-study-game-theory

11. Pan, Y., Pavel, L.: Games with coupled propagated constraints in optical networks with multi-link topologies. Automatica **45**(4), 871–880 (2009)
12. Salehisadaghiani, F., Pavel, L.: Distributed Nash equilibrium seeking: a gossip-based algorithm. Automatica **72**, 209–216 (2016)
13. Salehisadaghiani, F., Pavel, L.: Distributed Nash equilibrium seeking by gossip in games on graphs. In: 2016 IEEE 55th Conference on Decision and Control (CDC), pp. 6111–6116. IEEE (2016)
14. Salehisadaghiani, F., Pavel, L.: Nash equilibrium seeking with non-doubly stochastic communication weight matrix. arXiv preprint (2016). arXiv:1612.07179
15. Salehisadaghiani, F., Pavel, L.: A distributed Nash equilibrium seeking in networked graphical games. arXiv preprint (2017). arXiv:1703.09765
16. Salehisadaghiani, F., Pavel, L.: Distributed Nash equilibrium seeking via the alternating direction method of multipliers. IFAC World Congress (to appear, 2017)
17. Yin, H., Shanbhag, U.V., Mehta, P.G.: Nash equilibrium problems with scaled congestion costs and shared constraints. IEEE Trans. Autom. Control **56**(7), 1702–1708 (2011)
18. Zhu, M., Frazzoli, E.: Distributed robust adaptive equilibrium computation for generalized convex games. Automatica **63**, 82–91 (2016)

A Multitype Hawk and Dove Game

Aditya Aradhye[1], Eitan Altman[2,3,4](✉), and Rachid El-Azouzi[4]

[1] Chennai Mathematical Institute, Chennai, India
adityaaradhye@gmail.com
[2] Univ Côte d'Azur, Inria, Sophia Antipolis, France
Eitan.Altman@inria.fr
[3] LINCS, Paris, France
[4] LIA, University of Avignon, Avignon, France
rachid.elazouzi@univ-avignon.fr
http://www-sop.inria.fr/members/Eitan.Altman/

Abstract. We consider in this paper the Hawk-Dove game in which each of infinitely many individuals, involved with pairwise encounters with other individuals, can decide whether to act aggressively (Hawk) or peacefully (Dove). Each individual is characterized by its strength. The strength distribution among the population is assumed to be fixed and not to vary in time. If both individuals involved in an interaction are Hawks, there will be a fight, the result of which will be determined by the strength of each of the individuals involved. The larger the difference between the strength of the individuals is, the larger is the cost for the weaker player involved in the fight. Our goal is to study the influence of the parameters (such as the strength level distribution) on the equilibrium of the game. We show that for some parameters there exists a threshold equilibrium policy while for other parameters there is no equilibrium policy at all.

1 Introduction

Evolutionary games become a central tool for predicting and even design evolution in many field. The origin of evolutionary games come from biology where it was introduced by [15] to model conflicts among animals. It differs from classical game theory by (i) its focusing on the evolution dynamics of the fraction of members of the population that use a given strategy, and (2) in the notion of Evolutionary Stable Strategy (ESS, [15]) which includes robustness against a deviation of a whole (possibly small) fraction of the population who may wish to deviate (This is in contrast with the standard Nash equilibrium concept that only incorporates robustness against deviation of a single user). It became perhaps the most important mathematical tool for describing and modeling evolution since Darwin. Indeed, on the importance of the ESS for understanding the evolution of species, Dawkins writes in his book "The Selfish Gene" [18]: "we may come to look back on the invention of the ESS concept as one of the most important advances in evolutionary theory since Darwin." He further specifies: "Maynard Smith's concept of the ESS will enable us, for the first time, to see clearly how a

© ICST Institute for Computer Sciences, Social Informatics and Telecommunications Engineering 2017
L. Duan et al. (Eds.): GameNets 2017, LNICST 212, pp. 16–28, 2017.
DOI: 10.1007/978-3-319-67540-4_2

collection of independent selfish entities can come to resemble a single organized whole." Recently, however, evolutionary game theory has become of increased interest to social scientists [8]. In computer science, evolutionary game theory is appearing, some examples of applications can be found in multiple access protocols [16], multihoming [14] and resources competition in the Internet [20].

In this paper we focus on the classical evolutionary Hawk-Dove game which is one of the most studied examples in evolutionary games. The Hawk-Dove game is a model for determining the degree of aggressiveness in a society in which each individual can decide on whether to be a Hawk or a Dove. There are many pairwise interactions between individuals while competing over resources such as food. While Hawkish behavior benefits an individual who meets a Dove while contending over a resource, it has a cost since it is involved in more confrontations Hawk-Hawk which are more violent and in which chances of getting wounded are high. The objective of a game analysis is then to predict what fraction of the population would be aggressive at equilibrium as a function of the system's parameters.

In this paper we assume that the choice between Dove and Hawk only determines whether or not there would be a confrontation between the individuals. But the outcome of the conflict is determined by a parameter that is proper to each of the involved individual, which we call strength. It could be related to its size, or its weight. Each individual in a large population takes a decision on whether to act aggressively (Hawk) or not (Dove) based on its own strength. It is again involved in many pairwise encounters with other individuals randomly selected from a large population. The decision to act aggressively or not is taken without knowing what will be the strength of those individuals it would meet.

In evolutionary game literature, variants of the hawk-dove game exist. For example, in [11] a dynamic version of the hawk dove game is proposed. In this version, it is assumed that each player (animal) has a state that corresponds to its level of energy reserves. A strategy of a player specifies which action to take as a function of its state. Assuming that an animal must minimize its probability of dying, the authors established a new ESS according to which an animal plays a hawk if its energy reserves are below some critical value, and plays dove otherwise. Furthermore, any single mutant that adopts other strategy than the ESS would get a strictly lower fitness. In [19], the author considered a heterogeneous population composed of two groups of hawks and doves that have different fighting abilities and which are linked via migration. Assuming that migration occurs at a much faster time scale than the game dynamics (hawk dove game), the author studied the dynamics of the full population. Other interesting extended versions of the Hawk-Dove game are proposed in [3–7, 10, 12, 13].

Originated in biology, the hawk-dove game lends itself well to various networking problems such as power control or medium access control as well. In [2, 9], a semi-dynamic version of the hawk-dove game applied to power control is introduced. In this game, the aggressive behavior stands for transmitting at a high power level while the peaceful behavior is associated to transmitting at a low power level. Each mobile station (player) has a state that corresponds to its energy level. The action used by a player determines its immediate fitness and its future state. Moreover, it is assumed that a player can use only the same strategy during its lifetime. The goal of a player is to maximize its overall amount of data sent during its lifetime. The authors identified in this context a paradox in which the fraction of a population choosing the peaceful behavior at the evolutionarily stability decreases as the initial energy state of players increases. In [1], the authors applied the hawk dove game to congestion control where the aggressive behavior corresponds to using a high-speed TCP version to be used over the Internet. Another application of the hawk-dove game in the medium access control is considered in [17].

In this paper we allow the state space to be a continuum. After presenting the model in the next section, we identify in Sect. 3 conditions for an equilibrium with a threshold structure to exist, in which a individual behaves aggressively if it is stronger than some threshold. We then search in Sect. 4 for other equilibria and show that under some conditions, any equilibrium other than threshold does not exist at all. This is due to the fact that the state space is infinite and has thus not been observed in games with finite state spaces.

2   Model

We consider a Hawk and Dove game, in which individuals have pairwise interactions over resources (food). Two individuals that adopt a Dove behavior split the resource peacefully; we assume that the share of each individual depends on the strength of the individual as follows. The stronger individual receives a fraction α of the resource and the other one receives $1 - \alpha$ of it, where α is a constant between 0 and 1. If it meets a individual with an aggressive behavior (Hawk) then the whole resource is taken by the aggressive individual so that the Hawk gets one unit of fitness and the Dove none.

When two Hawks meet, there is a fight in which case the true identity determines the fitness of each player. We assume that the stronger individual receives one unit of fitness whereas the weaker one's utility is monotone increasing in its strength. Let x (and y) be the strength levels of the stronger (resp. weaker) individual. Then we assume that the fitness of the weaker individual is given by $-f(y, x - y)$ for some nonnegative f which is assumed to be decreasing in its first argument and increasing in its second one. Each individual is encountered with another individual chosen uniformly at random. Assume that the strength level in the population is distributed according to probability density function $\theta(x)$. This means the probability that any individual encounters a individual of strength between x to $x + dx$ is $\theta(x)dx$ for small dx. When stronger individual has strength x and weaker individual has strength $y(< x)$, then payoff matrices are as follows (Tables 1 and 2).

The probability that a individual of strength x meets another individual of strength exactly equal to x is zero. Hence the payoffs in that case do not affect the utility of the individuals.

We define P (as a function of x) to be the strategy of the individuals. $P(x)$ is the probability of playing Hawk when at strength level x (and hence probability

Table 1. Payoff for player with strength $x(> y)$

	H	D
H	1	1
D	0	α

Table 2. Payoff for player with strength $y(< x)$

	H	D
H	$-f(x, y - x)$	1
D	0	$1 - \alpha$

that it plays Dove is $1 - P(x)$). In general different individuals can play different strategies. But when a particular individual is encountered by a random individual, only thing that affects its utility is the probability with which it is encountered by a individual with strength y and strategy Hawk and individual with strength y and strategy Dove. If $h(y)$ is the probability with which the individual is encountered by a individual with strength y and playing Hawk, we can equivalently assume that all individuals with strength y are playing Hawk with probability $\frac{h(y)}{\theta(y)}$. Hence forward we shall assume all the individuals in the environment play the same strategy. Define the utility of the individual as $U(P', x, P)$, this is the expected utility that a individual (of strength x) gets when it uses strategy P' and the rest of the population uses strategy P. It is given by

$$U(P', x, P) = P'(x)U(H, x, P) + (1 - P'(x)) U(D, x, P)$$

where H is pure strategy of playing Hawk and D is pure strategy of playing Dove. We have,

$$U(H, x, P) = \int_0^x \theta(y) \, dy + \int_x^\infty \theta(y)P(y)(-f(x, y-x)) \, dy + \int_x^\infty \theta(y)(1-P(y)) \, dy$$

$$U(D, x, P) = \alpha \int_0^x \theta(y)(1 - P(y)) \, dy + (1 - \alpha) \int_x^\infty \theta(y)(1 - P(y)) \, dy$$

3 Threshold Strategy

Before studying the existence of Nash equilibrium and ESS, let us define the threshold strategy based on the strength level.

Definition 1. *We define a threshold strategy P by*

$$P(x) = \begin{cases} 0 & \text{if } x < L \\ 1 & \text{if } x > L \\ \text{any value} \in [0, 1] & \text{if } x = L \end{cases} \tag{1}$$

We call L as threshold value of this threshold strategy, and denote the threshold strategy as P_L.

Theorem 1. *If f is a bounded function, then there exists a threshold strategy (for $\alpha = 1/2$), such that if it is used by the individual and the population, it is Nash equilibrium. If the function f is strictly increasing in x, then this threshold strategy is also an ESS.*

Proof. Part 1 - Existence of threshold strategy which is Nash equilibrium.

We find the conditions on threshold value L so that its threshold strength is Nash equilibrium. Let P_L be a Nash equilibrium. This $U(P_L, x, P_L) \geq U(P, x, P_L)$ for every strategy P, and every strength level x. We have

$$P_L(x)U(H, x, P_L) + (1 - P_L(x))U(D, x, P_L) \geq P(x)U(H, x, P_L) + (1 - P(x))U(D, x, P_L) \quad (2)$$

Case 1 : $x > L$

For $x > L$, we have $P_L(x) = 1$, to satisfy (2), we must have
$U(P_L, x, P_L) \geq U(P, x, P_L)$
$U(H, x, P_L) \geq P(x)U(H, x, P_L) + (1 - P(x))U(D, x, P_L)$
$(1 - P(x))U(H, x, P_L) \geq (1 - P(x))U(D, x, P_L)$

Above inequality has to be satisfied by all strategies P, so it is necessary and sufficient that
$U(H, x, P_L) \geq U(D, x, P_L)$
$\int_0^x \theta(y)\,dy + \int_x^\infty \theta(y)(-f(x, y - x))\,dy \geq \frac{1}{2}\int_0^L \theta(y)\,dy$
$\int_0^x \theta(y)\,dy \geq \frac{1}{2}\int_0^L \theta(y) + \int_x^\infty \theta(y)f(x, y - x)\,dy$

LHS of the above statement is increasing in x and as f is decreasing in x, RHS is decreasing in x. So, it is necessary and sufficient that inequality is satisfied for $x = L$.

$$\int_0^L \theta(y)\,dy \geq \frac{1}{2}\int_0^L \theta(y) + \int_L^\infty \theta(y)f(L, y - L)\,dy$$

$$\frac{1}{2}\int_0^L \theta(y)\,dy \geq \int_L^\infty \theta(y)f(L, y - L)\,dy \quad (3)$$

Case 2 : $x < L$

For $x < L$, we have $P_L(x) = 0$, to satisfy (2), we must have
$U(P_L, x, P_L) \geq U(P, x, P_L)$
$U(D, x, P_L) \geq P(x)U(H, x, P_L) + (1 - P(x))U(D, x, P_L)$
$P(x)U(D, x, P_L) \geq P(x)U(H, x, P_L)$

It is necessary and sufficient to have $U(D, x, P_L) \geq U(H, x, P_L)$
$\int_0^L \theta(y)\,dy + \int_L^\infty \theta(y)(-f(x, y - x))\,dy \leq \frac{1}{2}\int_0^L \theta(y)\,dy$

$\frac{1}{2} \int_0^L \theta(y) \, dy \leq \int_L^\infty \theta(y) f(x, y - x) \, dy$

LHS of the above statement is constant in x and as f is decreasing in x, RHS is decreasing in x. So, it is necessary and sufficient that inequality is satisfied for $x = L$.

$\int_0^L \theta(y) \, dy \leq (1/2) \int_0^L \theta(y) + \int_L^\infty \theta(y) f(L, y - L) \, dy$

$$\frac{1}{2} \int_0^L \theta(y) \, dy \leq \int_L^\infty \theta(y) f(L, y - L) \, dy \tag{4}$$

(3) and (4) imply

$$\frac{1}{2} \int_0^L \theta(y) \, dy = \int_L^\infty \theta(y) f(L, y - L) \, dy \tag{5}$$

Above equation also tells us that $U(H, L, P_L) = U(D, L, P_L)$, so a player with strength L is indifferent in playing Hawk and Dove, so $P_L(L)$ can take any value between 0 and 1 and still P_L will be a Nash equilibrium. So, it is sufficient for L to satisfy the above equation for P_L to be Nash equilibrium.

LHS of (5) is increasing in L and RHS of (5) is decreasing in L. At $L = 0$, LHS takes value 0 and as L tends to ∞, LHS tends to $1/2$. If RHS is bounded, as f is a non negative, not identically zero bounded function, RHS takes a positive value at $L = 0$ and tends to 0 as L tends to ∞. Hence, (5) has unique solution. So, we have a threshold strategy which is Nash equilibrium whenever the RHS is bounded.

Part 2 -
For P_L to be ESS, for all strategies P other than P_L and for all x except maybe on a set of measure zero, at least one of the following must hold,

(1) $U(P_L, x, P_L) > U(P, x, P_L)$
(2) $U(P_L, x, P_L) = U(P, x, P_L)$ and $U(P_L, x, P) > U(P, x, P)$

Let P_L be the threshold strategy we get in Part 1 which is Nash equilibrium. Then we have, $U(H, L, P_L) = U(D, L, P_L)$

Also, as for $x \geq L$, $U(D, x, P_L) = \frac{1}{2} \int_0^L \theta(y) \, dy$ and

$U(H, x, P_L) = \int_0^x \theta(y) \, dy + \int_x^\infty \theta(y) P(y)(-f(x, y - x)) \, dy$, we have

$U(H, x, P_L) > U(H, L, P_L)$ for $x > L$ (Strict inequality because f is strictly increasing)

So for $x > L$, we have

$U(H, x, P_L) > U(H, L, P_L) = U(D, L, P_L) = U(D, x, P_L)$, which implies

$U(P_L, x, P_L) > U(P, x, P_L)$

Similarly, we can also show the same for $x < L$. This proves that P_L is also an ESS as L is set of measure zero.

Threshold equilibrium for general α: We now try to find if there exists a threshold Nash equilibrium strategy for general α. We assume the probability density function θ is differentiable and decreasing. We assume f is bounded and decreasing in x. It is also reasonable to assume that f is concave in the variable x (meaning $\frac{\delta^2}{\delta^2 x} f(x, y - x) \le 0$ for all x).

Case 1 : $x \ge L$

In this case, $P_L(x) = 1$. For P_L to be Nash equilibrium, we must have for any other strategy P and for all $x \ge L$, $U(P_L, x, P_L) \ge U(P, x, P_L)$

$\Leftrightarrow U(H, x, P_L) \ge P(x) U(H, x, P_L) + (1 - P(x)) U(D, x, P_L)$

$\Leftrightarrow (1 - P(x)) U(H, x, P_L) \ge (1 - P(x)) U(D, x, P_L)$

$\Leftrightarrow U(H, x, P_L) \ge U(D, x, P_L)$

$\Leftrightarrow \int_x^\infty \theta(y)(-f(x, y - x)) \, dy + \int_0^x \theta(y) \, dy \ge \alpha \int_0^L \theta(y) \, dy$

$\Leftrightarrow \int_0^x \theta(y) \, dy \ge \alpha \int_0^L \theta(y) + \int_x^\infty \theta(y) f(x, y - x) \, dy$

LHS of the above statement is increasing in x and as f is decreasing in x, RHS is decreasing in x. So, it is sufficient that equation is satisfied for $x = L$.

$$\int_0^L \theta(y) \, dy \ge \alpha \int_0^L \theta(y) + \int_L^\infty \theta(y) f(L, y - L) \, dy$$

$$(1 - \alpha) \int_0^L \theta(y) \, dy \ge \int_L^\infty \theta(y) f(L, y - L) \, dy \qquad (6)$$

Case 2 : $x < L$

In this case, $P_L(x) = 0$. For P_L to be Nash equilibrium, we must have for any other strategy P and for all $x \le L$,

$U(P_L, x, P_L) \ge U(P, x, P_L)$

$\Leftrightarrow U(D, x, P_L) \ge P(x) U(H, x, P_L) + (1 - P(x)) U(D, x, P_L)$

$\Leftrightarrow P(x) U(D, x, P_L) \ge P(x) U(H, x, P_L)$

$\Leftrightarrow U(D, x, P_L) \ge U(H, x, P_L)$

$\Leftrightarrow \alpha \int_0^x \theta(y) \, dy + (1 - \alpha) \int_x^L \theta(y) \, dy \ge \int_L^\infty \theta(y)(-f(x, y - x)) \, dy + \int_0^L \theta(y) \, dy$

$$\Leftrightarrow \alpha \int_0^L \theta(y) \, dy \le (2\alpha - 1) \int_0^x \theta(y) \, dy + \int_L^\infty \theta(y) f(x, y - x) \, dy \qquad (7)$$

Let $F(x)$ denote the RHS of (7). LHS is independent of x, so (7) holds for $x < L$ if and only if $F(x)$ satisfies the inequality $LHS \leq min_{0 \leq x \leq L} \{F(x)\}$.

$$F(x) = (2\alpha - 1) \int_0^x \theta(y) + \int_L^\infty \theta(y) f(x, y - x) \, dy$$

As θ is assumed differentiable, F is twice differentiable.

$$F'(x) = (2\alpha - 1)\theta(x) + \int_L^\infty \theta(y) \tfrac{\delta}{\delta x} f(x, y - x) \, dy$$

$$F''(x) = (2\alpha - 1)\theta'(x) + \int_L^\infty \theta(y) \tfrac{\delta^2}{\delta^2 x} f(x, y - x) \, dy$$

As f is concave in x and θ is decreasing, $F''(x) \leq 0 \; \forall x \leq L$

So, F takes minimum value either at 0 or L.

For this case, it is necessary and sufficient that (7) is satisfied by $x = 0$ and $x = L$. These with (6) are the necessary and sufficient conditions for L to Nash equilibrium threshold strategy. So, P_L is Nash equilibrium if and only if (1) $\alpha \int_0^L \theta(y) \, dy = F(L)$ and (2) $\alpha \int_0^L \theta(y) \, dy \leq F(0)$.

So there exists a threshold strength L such that P_L is Nash equilibrium if and only if there exists L satisfying both above equations, which when rearranged can be written as follows

$$(1 - \alpha) \int_0^L \theta(y) \, dy = \int_L^\infty \theta(y) f(L, y - L) \, dy \tag{8}$$

$$\alpha \leq \frac{1}{2} + \frac{\int_L^\infty \theta(y)[f(0, y) - f(L, y - L)] \, dy}{2 \int_0^L \theta(y) \, dy} \tag{9}$$

As we can see, this proves that for $\alpha = 1$, (8) cannot be satisfied, hence there is no solution. For $\alpha \leq \frac{1}{2}$, there always exists a Nash equilibria threshold strength. For $\frac{1}{2} < \alpha < 1$, existence of a Nash equilibrium threshold strategy depends upon whether the functions θ and f satisfy (9).

4   Other Equilibria

We would now try to find if there are other equlibria. For this section we shall assume the function f to be only dependent on the difference between the individual's and the opponent's strength levels (earlier f was dependent on the difference in strength levels and individual's strength level). We also assume $\alpha = 1/2$. Note that if at some strength level x, if the individual receives more utility by playing Hawk (Dove) then it will play Hawk (Dove) with full probability in the equilibrium.

Hence, $U(H, x, P) > U(D, x, P) \Rightarrow P(x) = 1$ for P to be equilibrium strategy. Similarly, $U(H, x, P) < U(D, x, P) \Rightarrow P(x) = 0$

Also, $P(x) = 1 \Rightarrow U(H, x, P) \geq U(D, x, P)$,

$P(x) = 0 \Rightarrow U(H, x, P) \leq U(D, x, P),$

$0 < P(x) < 1 \Rightarrow U(H, x, P) = U(D, x, P)$ for P to be equilibrium strategy

Lemma: Any equilibrium strategy is monotone. Formally for an equilibrium strategy P, for some x_1, if $U(H, x_1, P) > U(D, x_1, P)$, then $P(x) = 1 \; \forall x \geq x_1$, for some x_2, if $U(H, x_2, P) < U(D, x_2, P)$, then $P(x) = 0 \; \forall x \leq x_2$.

Proof:

$U(H, x, P) = \int_0^x \theta(y) \, dy + \int_x^\infty \theta(y) P(y)(-f(y - x)) \, dy + \int_x^\infty \theta(y)(1 - P(y)) \, dy$
After simplifying, we get $U(H, x, P) = 1 - \int_x^\infty \theta(y) P(y)(1 + f(y - x)) \, dy$

f is increasing in $y - x$, so f is decreasing in x. So the quantity $\int_x^\infty \theta(y) P(y)(1 + f(y - x)) \, dy$ is decreasing in x. So, $U(H, x, P)$ is increasing in x.

$U(D, x, P) = (1/2) \int_0^\infty \theta(y)(1 - P(y)) \, dy$ is independent of x. So, the quantity $U(H, x, P) - U(D, x, P)$ is increasing in x.

$U(H, x_1, P) > U(D, x_1, P) \rightarrow U(H, x_1, P) - U(D, x_1, P) > 0$

For $x \geq x_1$, $U(H, x, P) - U(D, x, P) \geq U(H, x_1, P) - U(D, x_1, P) > 0 \rightarrow P(x) = 1$.

By similar reasoning, second part also holds. This completes the proof of the lemma.

Let $A_P = \{x \, | U(H, x, P) > U(D, x, P)\}$, $B_P = \{x \, | U(H, x, P) = U(D, x, P)\}$, $C_P = \{x \, | U(H, x, P) < U(D, x, P)\}$.

Because of our lemma in this section, every element in B_P is greater than every element in A_P and less than every element in C_P.

Let $x_1 = \inf(B_P)$, $x_2 = \sup(B_P)$. Clearly, $x_1 \leq x_2$

So, we have $P(x) = 0 \; \forall x < x_1$ and $P(x) = 1 \; \forall x > x_2$

If $x_1 = x_2$, $P(x)$ can take non trivial value (value greater than 0 and less than 1) at the most one value (It may take nontrivial value at $x_1 (= x_2)$). So the strategy is a threshold strategy.

Now assume $x_1 < x_2$. For $x_1 < x < x_2, 0 < P(x) < 1$. So, we have $U(H, x, P) = U(D, x, P)$ on the whole interval (x_1, x_2). So, we can differentiate and equate the two sides.

For $x_1 < x < x_2$, the utilities are

$U(H, x, P) = \int_0^x \theta(y)\, dy + \int_x^{x_2} \theta(y)(-f(y-x))P(y)\, dy + \int_x^{x_2} \theta(y)(1 - P(y))\, dy + \int_{x_2}^\infty \theta(y)(-f(y-x))\, dy$

$U(D, x, P) = (1/2) \int_0^{x_1} \theta(y)\, dy + (1/2) \int_{x_1}^{x_2} \theta(y)(1 - P(y))\, dy$

$U(H, x, P) = U(D, x, P)$

$\Leftrightarrow \int_0^x \theta(y)\, dy + \int_x^{x_2} \theta(y)(-f(y-x))P(y)\, dy + \int_x^{x_2} \theta(y)(1 - P(y))\, dy + \int_{x_2}^\infty \theta(y)(-f(y-x))\, dy = (1/2) \int_0^{x_1} \theta(y)\, dy + (1/2) \int_{x_1}^{x_2} \theta(y)(1 - P(y))\, dy$

$\Leftrightarrow \int_0^{x_2} \theta(y)\, dy - \int_x^{x_2} \theta(y)(1 + f(y-x))P(y)\, dy + \int_{x_2}^\infty \theta(y)(-f(y-x))\, dy = (1/2) \int_0^{x_1} \theta(y)\, dy + (1/2) \int_{x_1}^{x_2} \theta(y)(1 - P(y))\, dy$

We have $U(H, x, P) = U(D, x, P)$ on the whole interval (x_1, x_2). So, we can differentiate (with respect to x) and equate the two sides. We get,

$-(\int_x^{x_2} \theta(y) \frac{\delta}{\delta x} f(y-x) P(y)\, dy) + \theta(x)(1 + f(0)) P(x) - (\int_{x_2}^\infty \theta(y) \frac{\delta}{\delta x} f(y-x)\, dy) = 0$

$\theta(x)(1 + f(0)) P(x) = \int_x^{x_2} \theta(y) \frac{\delta}{\delta x} f(y-x) P(y)\, dy + + \int_{x_2}^\infty \theta(y) \frac{\delta}{\delta x} f(y-x)\, dy$

Since f is increasing in $(y-x)$, it is decreasing in x. So, the quantity $\frac{\delta}{\delta x} f(y-x)$ is negative. Both the terms in the RHS are nonpositive, so RHS is nonpositive. But LHS is nonnegative, so for this equation to satisfy we must have $P(x) = 0 \ \forall x_1 < x < x_2$. Contradiction, since x such that $x_1 < x < x_2$ belongs to B_P, and hence $P(x) > 0$.

So, for $x_1 < x_2$, we do not have any solution. So, there does not exist any other equilibrium strategy other than threshold strategy.

5 Price of Stability

In this section, we study the inefficiency caused in the objective function by imposing the condition of Nash equilibrium for certain functions f. Objective function here for us is the average utility of all the players. This inefficiency is quantified by price of Stability. Formally it is defined as

Price of Stability $(PoS) = \frac{max_{P \in \mathcal{P}_e} AU(P)}{max_{P \in \mathcal{P}} AU(P)}$

where $AU(P)$ is average utility when the whole population plays strategy P, $\mathcal{P}$ is set of all strategies for the population, $\mathcal{P}_e$ is set of all equilibrium strategies for the population. Thus, PoS is the ratio of best you can get using the Nash equilibrium strategies and the overall best you can get. Note, PoS is always less than or equal to 1. We shall find PoS for some density functions θ and cost functions f. Apart from previous conditions on θ and f, we would generally put the restrictions on f that $f(x, 0) = 0$ and fix a constant, say $\frac{1}{2}$, we would generally want for fixed x, $f(x, y-x)$ tend to infinity or a constant at least $\frac{1}{2}$ as $y-x$ tend to infinity. We shall now calculate PoS for some examples with the restrictions defined.

Example 1. We consider the case where $\alpha = \frac{1}{2}$. Let $\theta(y) = \mu e^{-\mu y}$ and $f(x, y-x) = 1 - e^{-\mu(y-x)}$.

The value of threshold strength L can be calculated by solving the equation

$$\frac{1}{2}\int_0^L \theta(y)\, dy = \int_L^\infty \theta(y) f(L, y-L)\, dy$$

Solving, we get $e^{-\mu L} = \frac{1}{2}$ or $L = \frac{1}{\mu} \ln(2)$.

Thus, the utility for player with strength $x < L$ is $\frac{1}{2}\int_0^L \mu e^{-\mu y}\, dy = \frac{1}{2}(1 - e^{-\mu L}) = \frac{1}{4}$

For $x > L$ it is $\int_0^x \mu e^{-\mu y}\, dy - \int_x^\infty \mu^{-\mu y}(1 - e^{-\mu(y-x)})\, dy = 1 - \frac{3}{2}e^{-\mu x}$

$$\text{Average utility } AU = \int_0^L \mu e^{-\mu x}(\frac{1}{4})\, dx + \int_L^\infty \mu e^{-\mu x}(1 - \frac{3}{2}e^{-\mu x})\, dx$$

$$= \frac{1}{4}(1 - e^{-\mu L}) + e^{-\mu L} - (\frac{3}{2})(\frac{1}{2})e^{-2\mu L}$$

$$= \frac{1}{4}(1 - \frac{1}{2}) + \frac{1}{2} - (\frac{3}{2})(\frac{1}{2})\frac{1}{4} = \frac{7}{16}$$

As this is unique Nash equilibrium, it is best Nash equilibrium. It can be clearly seen that when strategies are not restricted to Nash equilibrium, the average payoff is maximized when all players play Dove, the average payoff in this case is $\frac{1}{2}$. So, PoS for this game $\frac{7}{8}$.

Example 2. Consider the same α and θ, but $f(x, y-x) = 1 - e^{-c\mu(y-x)}$ for some constant > 0. By same calculations, we can check that

$e^{-\mu L} = \frac{c+1}{3c+1}$ and $AU = \frac{1}{2} - \frac{c(c+1)}{2(3c+1)^2}$.

We cannot have $c = 0$ (in that case, f(x, y-x) = 0 which is not allowed), but any $c > 0$ is allowed. As $c \downarrow 0$, $AU \to \frac{1}{2}$ and so $PoS \to 1$. So, for any small $\varepsilon > 0$, we can create a game by choosing proper value of c such that PoS of this game is $1 - \varepsilon$.

Example 3. Same α and θ, $f(x, y - x) = e^{k\mu(y-x)}$ where $k \geq 0$ is a constant. We can find threshold strength only for $k < 1$. So, for $k \geq 1$, there is no Nash equilibrium strategy, so by definition, PoS is zero. For $k < 1$, we get $e^{-\mu L} = \frac{1-k}{3-k}$ and $AU = \frac{1}{2} - \frac{1-k}{2(3-k)^2}$. As $k \uparrow 1$, $AU \to \frac{1}{2}$ and so $PoS \to 1$. So, again for any small $\varepsilon > 0$, we can create a game by choosing proper value of k such that PoS of this game is $1 - \varepsilon$.

References

1. Altman, E., El-Azouzi, R., Hayel, Y., Tembine, H.: The evolution of transport protocols: an evolutionary game perspective. Comput. Netw. **53**(10), 1751–1759 (2009)
2. Altman, E., Fiems, D., Haddad, M., Gaillard, J.: Semi-dynamic hawk and dove game, applied to power control. In: Proceedings IEEE INFOCOM, pp. 2771–2775, March 2012
3. Altman, E., Gaillard, J., Haddad, M., Wiecek, P.: Dynamic hawk and dove games within flocks of birds. In: Hart, E., Timmis, J., Mitchell, P., Nakamo, T., Dabiri, F. (eds.) BIONETICS 2011. LNICST, vol. 103, pp. 115–124. Springer, Heidelberg (2012). doi:10.1007/978-3-642-32711-7_9
4. Altman, E., Brunetti, I.: Revisiting evolutionary game theory. In: IEEE Conference on Decision and Control (CDC) 2013, pp. 1842–1847 (2013)
5. Auger, P., Pontier, D.: Fast game dynamics coupled to slow population dynamics: a single population with hawk-dove strategies. Math. Comput. Modell. **27**(4), 81–88 (1998)
6. Ben Khalifa, N., ElAzouzi, R., Hayel, Y.: Evolutionary games in interacting communities. Dyn. Games Appl. J. **7**, 131–156 (2017)
7. Cressman, R.: The Stability Concept of Evolutionary Game Theory. Springer, Heidelberg (1992)
8. Friedman, E., Henderson, S.: Fairness and efficiency in processor sharing protocols to minimize sojourn times. In: Proceedings of ACM SIGMETRICS, pp. 229–337 (2003)
9. Haddad, M., Altman, E., Fiems, D., Gaillard, J.: Paradoxes in semi-dynamic evolutionary power control game: when intuition fools you!. IEEE Trans. Wirel. Commun. **12**(11), 5728–5739 (2013)
10. Hayel, Y., Belmega, E.V., Altman, E.: Hawks and doves in a dynamic framework. Dyn. Games Appl. **3**(1), 24–37 (2012)
11. Houston, A.I., McNamara, J.M.: Fighting for food: a dynamic version of the hawk-dove game. Evol. Ecol. **2**(1), 51–64 (1988)
12. McNamara, J.M., Houston, A.I.: If animals know their own fighting ability, the evolutionarily stable level of fighting is reduced. J. Theor. Biol. **232**(1), 1–6 (2005)
13. McNamara, J.M., Merad, S., Collins, E.J.: The hawk-dove game as an average-cost problem. Adv. Appl. Probab. **23**(4), 667–682 (1991)
14. Shakkottai, S., Altman, E., Kumar, A.: The case for non-cooperative multihoming of users to access points in IEEE 802.11 WLANs. In: IEEE Infocom, Barcelona, Spain (2006)
15. Smith, M.: Game theory and the evolution of fighting. In: Maynard Smith, J. (ed.) On Evolution, pp. 8–28. Edinburgh University Press (1972)
16. Kumar, A., Shakkottai, S., Altman, E.: Evolutionary power control games in wireless networks. J. Sel. Areas Commun. **25**(6), 1207–1215 (2007)

17. Tembine, H., Altman, E., El-Azouzi, R.: Delayed evolutionary game dynamics applied to medium access control. In: Proceedings of the IEEE 4th International Conference on Mobile Adhoc and Sensor Systems, Pisa, Italy, pp. 1–6, October 2007
18. Wright, C.C.: An explanation for some aspects of the behaviour of congested road traffic in terms of a simple model. Transp. Res. **9**(5), 267–273 (1975)
19. Yearsley, J.: Hawks and doves in heterogeneous environments. Math. Comput. Modell. **27**(4), 99–108 (1998)
20. Zheng, Y., Feng, Z.: Evolutionary game and resources competition in the internet. In: Modern Communication Technologies 2001. SIBCOM-2001. The IEEE-Siberian Workshop of Students and Young Researchers, pp. 51–54 (2001)

Assortative Mixing Equilibria
in Social Network Games

Chen Avin[1], Hadassa Daltrophe[1]([⊠]), Zvi Lotker[1], and David Peleg[2]

[1] Ben Gurion University of the Negev, Beer Sheva, Israel
avin@cse.bgu.ac.il, hd@cs.bgu.ac.il, zvilo@bgu.ac.il
[2] Weizmann Institute of Science, Rehovot, Israel
david.peleg@weizmann.ac.il

Abstract. It is known that individuals in social networks tend to exhibit *homophily* (a.k.a. *assortative mixing*) in their social ties, which implies that they prefer bonding with others of their own kind. But what are the reasons for this phenomenon? Is it that such relations are more convenient and easier to maintain? Or are there also some more tangible benefits to be gained from this collective behaviour?

The current work takes a game-theoretic perspective on this phenomenon, and studies the conditions under which different assortative mixing strategies lead to equilibrium in an evolving social network. We focus on a biased preferential attachment model where the strategy of each group (e.g., political or social minority) determines the level of bias of its members toward other group members and non-members. Our first result is that if the utility function that the group attempts to maximize is the *degree centrality* of the group, interpreted as the sum of degrees of the group members in the network, then the only strategy achieving Nash equilibrium is a perfect homophily, which implies that cooperation with other groups is harmful to this utility function. A second, and perhaps more surprising, result is that if a reward for inter-group cooperation is added to the utility function (e.g., externally enforced by an authority as a regulation), then there are only two possible equilibria, namely, *perfect homophily* or *perfect heterophily*, and it is possible to characterize their feasibility spaces. Interestingly, these results hold regardless of the minority-majority ratio in the population.

We believe that these results, as well as the game-theoretic perspective presented herein, may contribute to a better understanding of the forces that shape the groups and communities of our society.

Keywords: Social networks · Homophily · Game theory

1 Introduction

Homophily (lit. "love of the same") [15], also known as *assortative mixing* [17], is a prevalent and well documented phenomenon in social networks [16]; in making

---

Supported in part by a grant of the Israel Science Foundation (1549/13).

© ICST Institute for Computer Sciences, Social Informatics and Telecommunications Engineering 2017
L. Duan et al. (Eds.): GameNets 2017, LNICST 212, pp. 29–39, 2017.
DOI: 10.1007/978-3-319-67540-4_3

their social ties, people often prefer to connect with other individuals of similar characteristics, such as nationality, race, gender, age, religion, education or
profession.

Homophily has many important consequences, both on the structure of the
social network (e.g., the formation of communities) and on the behaviors and
opportunities of participants in it, for example on the welfare of individuals [12]
and on the diffusion patterns of information in the network [13]. It is therefore
interesting to explore the reasons for this phenomenon. Clearly, one natural
reason is that relationship with similar individuals may be more convenient and
easier to maintain. But are there also some more tangible benefits to be gained
from this collective behaviour of sub-populations in the network?

To better understand homophily, we take a different perspective on this phenomenon and study it through a strategic, game-theoretic prism. We investigate the conditions under which different assortative (and disassortative) mixing
strategies lead to equilibrium in an *evolving social network* game.

To model the network evolution, we use a variant of the classical *preferential attachment* model [4], which incorporates a heterogeneous population and
assortative mixing patterns for the sub-populations. This model, known as *biased
preferential attachment* (BPA) [3], maintains the "rich get richer" property, but
additionally enables different mixing patterns (including perfect homophily and
heterophily) between sub-populations, by using rejection sampling.

In this paper, we modify this model by turning it into a game. Each sub-
population is represented as a player who can choose its mixing pattern as a
strategy. The *utility function* (or *payoff*) of a player is a result of its population's (expected) properties in the BPA model. A *strategy profile* (describing the
strategies of both players) attains a *Nash equilibrium* for the game if no player
can do better by unilaterally changing its own strategy.

Obviously, the result of the game depends on the players' utility functions. In
the current study we take an initial step and study two natural utility functions.
In the first, we consider the payoff to be the total power of the group, that is,
the sum of degrees of all group members. In this case we prove that there is a
unique stable Nash equilibrium which is the *perfect homophily* profile, namely,
cooperation with other groups is harmful to this utility function. We stress that
while there are other strategy profiles, like the unbiased profile, that guarantee
the same total power to the groups, those profiles do not yield Nash equilibrium.

Since perfect homophily results in complete segregation of the sub-
populations, we consider a second utility function based on a linear combination between the total power of the group and the number of *cross-population*
links (i.e., the size of the population cut). In particular, the utility is taken to
be γ times the total power of the group plus $1 - \gamma$ times the population cut
size, for some *weight factor* $0 \leq \gamma \leq 1$. Such a utility can be viewed as a rule
(or a law) imposed by a regulator to encourage cooperation between the two
sub-populations. At a first glance, this utility seems to lead to different Nash
equilibria for different γ values. Somewhat surprisingly, we show that only two
possible equilibria may emerge. For $\gamma > 1/2$, the *perfect homophily* profile is
the unique Nash equilibrium, and for $\gamma < 1/2$, the *heterophily* profile is the

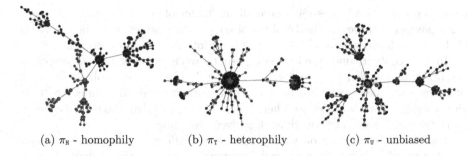

(a) π_{H} - homophily   (b) π_{T} - heterophily   (c) π_{U} - unbiased

Fig. 1. Examples of the Biased Preferential Attachment (BPA) model with various parameter settings. All examples depict a 200-vertex bi-populated network generated by our BPA model starting from a single edge connecting a blue and a red vertex and 30% red nodes (with vertex size proportional to its degree). (Color figure online)

unique Nash equilibrium. For $\gamma = 1/2$, both profiles yield a Nash equilibrium, but only the perfect homophily yields a stable equilibrium. (Note, by the way, that all our results are independent of the ratio r between the sizes of the two sub-populations.)

What may we learn from these results? A first, quite intuitive, lesson is that if the payoff includes benefits for heterophilic edges, then the game can move away from the perfect homophily equilibrium. But, within the natural utility function we study, if the game moves away from the homophily equilibrium, then it must reach a perfect heterophily equilibrium. Both of these equilibria may appear to be too "radical" from a social capital perspective, which may find it desirable to maintain some balance in-between the two extremes, i.e., preserve the internal structure of both sub-populations as well as form significant cross-population links between the two sub-populations. This leaves us with some interesting follow-up research directions: what 'mechanism design' rules can a regulator employ in order to have a more fine-grained control on the equilibrium? what happens in a system with more than two sub-populations? how do the equilibria behave? We leave these questions for future work; we believe that taking the game theoretic perspective on evolving social network models for heterogenous populations is an important tool in understanding homophily, as shown in this initial model.

Due to space limitations, we provide only an outline of our proofs. The interested reader is referred to [2] for details.

2   Related Work

Game theory provides a natural framework for modeling selfish interests and the networks they generate [1,18]. While many studies (see [11] for a comprehensive survey) focus on *local* network formation games, others (e.g., [7]) model the players as making *global* structural decisions. In this paper we define a game that features a mixture of both local and global characteristics. This situation is close to cooperative games [5], where all the nodes of the same group have the same payment. However, the key idea of cooperative games is to choose which coalitions to form, whereas here the partition into groups is predefined.

In this context, one should distinguish between *network formation games* [11,14,18] and *evolving network games* (e.g., [6]). The former involve a fixed set of nodes, with the connections between them changing over time. In contrast, in the evolving network model used herein, the nodes and edges are both dynamic, and new nodes join the network as it evolves over time.

Based on the assumption that people have tendency to copy the decisions of other people, we suggest a network construction process that follows the well known preferential attachment model [4] with an additional phase to incorporate the mixing parameter [3]. However, related studies in the economics literature examine different procedures to model the social network formation. The studies of [8,10] assume that individuals are randomly paired with other members of the population and then match assortatively. Another model, presented at [6], suggests two-phase attachments. The nodes first choose their neighbors with a bias towards their own type and then make an unbiased choice of neighbors from among the neighbors of their biased neighbors. While the models of [10,14] and others assume that a connecting edge between a pair of nodes is fixed by using bilateral agreement, in our model the matching choice is somewhat ambiguous. The rejection of a proposed connection can be interpreted as either decided by one of the parties unilaterally or accepted by a bilateral agreement.

One of the main themes of this paper is studying the homophily phenomenon and its influence on minority-majority groups. McPherson et al. [16] give an overview of research on homophily and survey a variety of properties and how they lead to particular patterns in bonding. While some studies (e.g., [3,8,9]) model homophily as ranging over a spectrum between perfect homophily and unbiased society, we have followed [6,10], which also allow disassortative matching.

Currarini, Jackson and Pin [8] examine friendship patterns in a representative sample of U.S. high schools and build a model of friendship formation based on empirical data. They report that all groups are biased towards same-type friendship relative to demographics, but different homophilic patterns emerge as a function of the group size; while homophily is essentially absent for groups that comprise very small or very large fractions of their school, it is significant for groups that comprise a middle-ranged fraction. In [10] it is also claimed that the majority group has greater tendency to homophily. In contrast, we have presented independence between the size of the group and the mixing pattern. Namely, the majority-minority parameter r does not influence the attained equilibria. This inconsistency can be explained by the different construction of the network ([8,10] assume random matching with biased agreement as mentioned above), or perhaps by the simplicity of our model and the fact that it involves only two groups.

3   Network and Game Model

Our network model is an extension of the bi-populated biased preferential attachment (BPA) model [3]. We use this model as the basis to an *evolving heterogeneous network game*. We start by describing the network model.

3.1   Biased Preferential Attachment Model

The *biased preferential attachment model*[1] (BPA) [3] is a bi-populated preferential attachment model obtained by applying the classical preferential attachment model [4] to a bi-populated minority-majority network augmented with homophily.

Definition 1 (BPA Model, $BPA(n, r, \pi)$). The model describes a bi-populated random evolving network with red and blue vertices, where n is the total number of nodes, r is the arrival rate of the red vertices and π is the *mixing matrix*. Denote the social network at time t by $G_t = (V_t, E_t)$, where V_t and E_t, respectively, are the sets of vertices and edges in the network at time t, and let $d_t(v)$ denote the degree of vertex v at time t. The process starts with an arbitrary initial bi-populated (red-blue) connected network G_0 with n_0 vertices and m_0 edges. For simplicity we hereafter assume that G_0 consists of one blue and one red vertex connected by an edge, but this assumption can be removed. This initial network evolves in n time steps as follows. In every time step t, a new vertex v enters the network. The arrival rate of the red nodes is denoted by $0 < r < 1$, i.e., the new vertex v is red with probability r and blue with probability $1 - r$.

In the first stage, v selects a *tentative* neighbor u at random by preferential attachment, i.e., with probability proportional to u's degree at time t,

$$\mathbb{P}[u \text{ is chosen}] = d_t(u) / \sum_{w \in V_t} d_t(w).$$

The second stage employs a 2×2 stochastic *mixing matrix*, π, composed of the stochastic *homophily vectors* of each player, π_R, π_B, i.e.,

$$\pi = \begin{pmatrix} \pi_R \\ \pi_B \end{pmatrix} = \begin{pmatrix} \rho_R & 1 - \rho_R \\ 1 - \rho_B & \rho_B \end{pmatrix}.$$

Letting $x \in \{R, B\}$ be v's color, the edge (v, u) is inserted into the graph with probability ρ_x when u's color is also x. If the colors differ, then the edge is inserted with probability $1 - \rho_x$. If the edge is rejected (i.e., is not inserted into the graph), then the two-stage procedure is restarted. This process is repeated until some edge $\{v, u\}$ has been inserted. Thus in each time step, one new vertex and one new edge are added to the existing graph.

---

[1] In fact, here we extend the model of [3] to allow heterophily.

Note that the mixing matrix π describes the degree of *segregation* (incorporated by using rejection sampling) of the system. In particular, using the *perfect homophily* matrix $\pi_H = \begin{pmatrix} H_R \\ H_B \end{pmatrix} = \begin{pmatrix} 1 & 0 \\ 0 & 1 \end{pmatrix}$, all added edges connect vertex pairs of the same color. At the other extreme, using the perfect *heterophily* matrix $\pi_T = \begin{pmatrix} T_R \\ T_B \end{pmatrix} = \begin{pmatrix} 0 & 1 \\ 1 & 0 \end{pmatrix}$, all added edges connect vertex pairs with different color. Similarly, using the *unbiased* strategy matrix $\pi_U = \begin{pmatrix} U_R \\ U_B \end{pmatrix} = \begin{pmatrix} .5 & .5 \\ .5 & .5 \end{pmatrix}$, edges are connected independently of the node colors. For intermediate values $0 < \rho_R, \rho_B < 1$, the players show a tendency to favor one kind of interaction over another. When $\rho_R, \rho_B > 0.5$, the players tend to be homophilic, and when $\rho_R, \rho_B < 0.5$, the players tend to be heterophilic. Figure 1 presents three examples of parameter settings for the BPA model on a 200-vertex bi-populated social network with $r = 0.3$ (30% red nodes), using π_H, π_T and π_U.

3.2   Evolving Heterogeneous Network Games

We now define the *evolving heterogeneous* EH (t, r, π, γ) network game (EH *game*, for short) between the two sub-populations. The game is played between two players, the red player R and the blue player B. (Note that we occasionally use R and B to denote either the *color*, the corresponding *set* of nodes, or the corresponding *player*. The exact meaning will be clear from the context.)

Assume r and G_0 are given to the players. Each player $X \in \{R, B\}$ can now choose its *strategy vector* as a mixing vector π_X in the mixing matrix π. Then the network evolves according the biased preferential attachment model BPA(t, r, π).

Let $n_t(R)$ and $n_t(B)$, respectively, denote the number of red and blue nodes at time $t > 0$, where $n_t = n_t(R) + n_t(B) = n_0 + t$. Denote by $d_t(R)$ (respectively, $d_t(B)$) the sum of degrees of the red (resp., blue) vertices present in the system at time $t \geq 0$. Altogether, the number of edges in the network at time t is $m_t = m_0 + t$, where $d_t(R) + d_t(B) = 2m_t$.

Let $C(G_t)$ denote the *cut* of the graph G_t defined by the red-blue partition of V_t, i.e., the set of edges that have one endpoint in R and the other in B. Formally,

$$C(G_t) = \{(u, v) \in E_t \mid u \in R, v \in B\}.$$

Let $\phi(G_t) = |C(G_t)|$ denote the size of the cut.

In our game, the *payoff* of each player is a combination of two quantities: the *total power* of its sub-population (namely, its expected sum of degrees), and the expected *cut size* $\phi(G)$. Observe that these quantities pull in opposite directions, hence they are balanced using a parameter $0 \leq \gamma \leq 1$ that will serve as a *weighting factor* for the *utility function* of the game. The parameter γ can be viewed as set by a regulator to enforce cooperation between sub-populations. Formally, the payoffs (utilities) of the players R and B at time t are

$$U_t^\gamma(\mathrm{R}) = \gamma\frac{d_t(\mathrm{R})}{d_t} + (1-\gamma)\frac{\phi_t}{2m_t} = \frac{1}{d_t}\Big(\gamma d_t(\mathrm{R}) + (1-\gamma)\phi_t\Big),$$
$$U_t^\gamma(\mathrm{B}) = \gamma\frac{d_t(\mathrm{B})}{d_t} + (1-\gamma)\frac{\phi_t}{2m_t} = \frac{1}{d_t}\Big(\gamma d_t(\mathrm{B}) + (1-\gamma)\phi_t\Big).$$

A strategy profile π is a *Nash equilibrium* for the game EH (t, r, π, γ) if no player $X \in \{R, B\}$ can do better by unilaterally changing its own strategy π_X. A Nash equilibrium for the game EH (t, r, π, γ) is *stable* if a small change in π for one player leads to a situation where two conditions hold: (i) the player who did not change has no better strategy in the new circumstance, and (ii) the player who did change is now playing with a strictly worse strategy. If both conditions are met, then the player who changed its π will return immediately to the Nash equilibrium, hence the equilibrium is *stable*. If condition (i) does not hold (but condition (ii) does), then the equilibrium is *unstable*.

4   Degree Maximization Game

Before studying the behavior of the general evolving heterogeneous network game, let us consider the solution of the game in the basic case where $\gamma = 1$ for every t, i.e., each player's utility depends only on the expected sum of degrees.

An urn process. The biased preferential attachment BPA(n, r, π) process can also be interpreted as a Polya's urn process, where each new edge added to the graph corresponds to two new balls added to the urn, one for each endpoint, and the balls are colored by the color of the corresponding vertices. In this interpretation, a time step of the original evolving network process corresponds to the arrival of a new ball x (which is red with probability r and blue with probability $1 - r$), and in the ensuing procedure, we choose an existing ball y from the urn uniformly at random; now, if x is of the same (respectively, different) color $x \in \mathrm{R}, \mathrm{B}$ as y, then with probability ρ_x (resp., $1 - \rho_x$) we add to the urn both x and a second copy of y (corresponding to the two endpoints of the added edge), and with probability $1 - \rho_x$ (resp., ρ_x) we reject the choice of y and repeat the experiment, i.e., choose another existing ball y' from the urn uniformly at random. This is repeated until the choice of y is not rejected. Hence the arrival of each new ball x results in the addition of exactly two new balls to the urn, namely, x and a copy of some existing ball y.

The key observation is that to analyze the expected fraction of the red balls in the urn at time t, there is no need to keep track of the degrees of individual vertices in the corresponding process of evolving network; the sum of degrees of all red vertices, $d_t(\mathrm{R})$, is exactly the number of red balls in the urn. Noting that exactly two balls join the system in each time step, we have

$$d_t(\mathrm{R}) + d_t(\mathrm{B}) = d_t = 2t + n_0 = 2(t + 1).$$

Note that while $d_t(\mathrm{R})$ and $d_t(\mathrm{B})$ are random variables, d_t is not.

Convergence of expectations. Let $\alpha_t = d_t(\mathrm{R})/d_t$ be a random variable denoting the fraction of red balls in the system at time t. Given the mixing matrix π, we claim that the process will converge to a ratio of α red balls in the system (as a function of π). More formally, we claim that, regardless of the starting condition, there exists a limit $\alpha = \lim_{t \to \infty} \mathbb{E}[\alpha_t]$.

Lemma 1. $\mathbb{E}[\alpha_{t+1} \mid \alpha_t] = \alpha_t + \dfrac{F(\alpha_t) - \alpha_t}{t + 2}$, where

$$F(x) = \frac{1}{2}\left(1 + \frac{\rho_B(-1+r)(-1+\alpha)}{-\alpha + \rho_B(-1+2\alpha)} + \frac{r\rho_R\alpha}{1 - \alpha + \rho_R(-1+2\alpha)}\right).$$

Lemma 2. The function $F(x)$ has the following properties:

1. $F(x)$ is monotonically increasing.
2. $F(x)$ has exactly one fixed point, $\alpha \in [0, 1]$.
3. The image of the unit interval by $F(x)$ is contained in the unit interval: $F([0,1]) = \left[\frac{r}{2}, \frac{1+r}{2}\right] \subset [0, 1]$.
4. If $x < \alpha$ then $x < F(x) < \alpha$ and if $x > \alpha$ then $x > F(x) > \alpha$.

Assume w.l.o.g. that $\alpha_t < \alpha$. By Lemma 2 $\alpha_t < F(\alpha_t) < \alpha$, so by Lemma 1 $\alpha_t < \mathbb{E}[\alpha_{t+1} \mid \alpha_t] < \alpha$. Taking expectations, we get that $\mathbb{E}[\alpha_t] < \mathbb{E}[\alpha_{t+1}] < \mathbb{E}[\alpha] = \alpha$. We have thus shown that the expected value of α_t converges to the fixed point α of $F(x)$. We have thus established the following.

Theorem 1. Given the rate r of red nodes and the mixing matrix π, for any initial graph, as t tends to infinity, the expected fraction of red balls, $\mathbb{E}[\alpha_t]$, converges to the unique real $\alpha \in (0, 1)$ satisfying the equation $F(\alpha) = \alpha$, or

$$2\alpha = 1 + \frac{\rho_B(-1+r)(-1+\alpha)}{-\alpha + \rho_B(-1+2\alpha)} + \frac{r\rho_R\alpha}{1 - \alpha + \rho_R(-1+2\alpha)}.$$

Hence the limit α is the solution of the cubic equation

$$(2 - 4\rho_B - 4\rho_R + 8\rho_B\rho_R)\alpha^3 + (-3 + 7\rho_B + \rho_B r + 4\rho_R - 10\rho_B\rho_R + r\rho_R - 4\rho_B r\rho_R)\alpha^2$$
$$+ (1 - 3\rho_B - 2\rho_B r - \rho_R + 3\rho_B\rho_R + 4\rho_B r\rho_R)\alpha + \rho_B r - \rho_B r\rho_R = 0.$$

Note that this limit is independent of the initial values d_0 and α_0 of the system.

Existence of a Nash Equilibrium. Having shown that for any given strategy profile π the expected fraction of red node degrees converges to α, we examine the influence of the different strategies on the utility functions.

Lemma 3. The limit α and $\mathbb{E}[\alpha_t]$ are monotone in the mixing matrix entries, i.e., both increase with increasing ρ_R and decrease with increasing ρ_B.

Given the utility functions $U_t^1(R) = d_t(R)$ and $U_t^1(B) = d_t(B)$, each player can choose its row in the mixing matrix π. By Theorem 1 we get that $U_{t \to \infty}^1(R) = d_t\alpha$ and $U_{t \to \infty}^1(B) = d_t(1 - \alpha)$. Lemma 3 implies that the red and blue players maximize their utility by increasing ρ_R and ρ_B, respectively. Hence, the homophily strategy profile π_H is strictly dominant for both players. The same applies for $t < \infty$.

Theorem 2. The homophily strategy profile π_H is a unique Nash equilibrium for the game $\mathrm{EH}(t, r, \pi, \gamma = 1)$.

5   Utilitiy Maximization Game

The evolving heterogeneous network game EH (t, r, π, γ) for a bi-populated network consists of two contrasting ingredients, the expected sum of degrees $d(\cdot)$ and the cut size $\phi(G)$. The following theorem expresses the impact of these forces on the system as a function of the weighting factor γ.

Theorem 3. *Consider the evolving network game* EH (t, r, π, γ) *for* $0 < r < 1$.

1. *For* $\gamma > 1/2$, *the homophily strategy profile* π_H *is a unique Nash equilibrium.*
2. *For* $\gamma < 1/2$, *the heterophily strategy profile* π_T *is a unique Nash equilibrium.*
3. *For* $\gamma = 1/2$, *the only two Nash equilibria are* π_H *and* π_T. *The homophily strategy profile* π_H *is a stable Nash equilibrium, while the heterophily strategy profile* π_T *is an unstable Nash equilibrium.*

Sketch of proof. Given that the new vertex at time $t + 1$ is blue, the probability P_{BB} that it attaches to a blue vertex satisfies

$$P_{BB}(\alpha_t) = (1 - \alpha_t)\rho_B + \alpha_t \rho_B P_{BB}(\alpha_t) + (1 - \alpha_t)\rho_B P_{BB}(\alpha_t),$$

hence $P_{BB}(\alpha_t) = \frac{\rho_B - \rho_B \alpha_t}{\rho_B + \alpha_t - 2\rho_B \alpha_t}$. Similarly, when the new vertex at time $t + 1$ is red, the probability that it attaches to a red vertex is $P_{RR}(\alpha_t) = \frac{\rho_R \alpha_t}{1 - \alpha_t + \rho_R(1 - 2\alpha_t)}$.

Let $N_t(x)$ and $M_t(x)$ be random variables denoting, respectively, the number of *new* red balls and cut edges at time t. We have $d_t(R) = d_0(R) + \sum_{i=1}^{t} N_t(\alpha_{i-1})$ and $\phi(G_t) = \phi(G_0) + \sum_{i=1}^{t} M_i(\alpha_{i-1})$. Define the *potential function* of the red player, denoted Δ_R, as the expected increment of its utility at step t. Then

$$\Delta_R = \mathbb{E}\left[U_{t+1}^\gamma(R) - U_t^\gamma(R) \mid \alpha\right] = \mathbb{E}\left[\gamma N_{t+1}(\alpha) + (1 - \gamma)M_{t+1}(\alpha)\right]$$
$$= \gamma(1 - (1 - r)P_{BB}(\alpha) + rP_{RR}(\alpha)) + (1 - \gamma)(1 - ((1 - r)P_{BB}(\alpha) + rP_{RR}(\alpha)))$$
$$= 1 - (1 - r)P_{BB}(\alpha) + r(2\gamma - 1)P_{RR}(\alpha).$$

Similar considerations imply that the potential function of the blue player is

$$\Delta_B = 1 - rP_{RR}(\alpha) + (1 - r)(2\gamma - 1)P_{BB}(\alpha).$$

The theorem follows by inspecting the value of the potential functions Δ_R and Δ_B for every γ and using Lemma 3 (for the monotonicity of $P_{RR}(\alpha)$ and $P_{BB}(\alpha)$ with the entries of the mixing matrix). $\qquad\square$

6   Discussion

This work investigates the assortative mixing phenomenon using a game theory perspective. Given some predefined rules related to the probability of connecting to other node, each player is allowed to determine its strategy in order to maximize its payoff. First we used a utility function that captures degree centrality, and showed that the expected sum of degrees and its limit are monotonically increasing with the homophily tendency. This directly implies that the homophily strategy is the unique Nash equilibrium. In this context, it will be interesting to use different centrality measures (such as PageRank, betweenness, etc.) and examine their influence on the equilibria. Next we enhanced the utility function to give positive payoff for both the degree and the cut. The results we have presented show a phase transition in the strategy as a function the weight γ. A small fluctuation in γ might cause extreme changes in the preference of the players, i.e., from perfect homophily to perfect heterophily (or vice versa); the intermediate strategies are never in equilibrium. This result is independent of the fraction of the sub-population size in the population. Generalizing the model to more than two sub-populations or reformulating the utility function may shape the strategy function differently.

An interesting outcome of the above is the possibility that setting a rule (or a law) by a regulator to encourage cooperation between the two sub-populations will play as a remedial strategy to achieve equal opportunities. This observation is remarkable since, in contrast to the usual affirmative action approach, this attitude does not discriminate any individual, but at the same time, it promises a fair representation of the different sub-populations and even a way for breaking the glass ceiling [3] that some minority sub-populations suffer from. We leave this direction for further work.

References

1. Aumann, R., Myerson, R.: Endogenous formation of links between players and coalitions: an application of the shapley value. In: The Shapley Value, pp. 175–191 (1988)
2. Avin, C., Daltrophe, H., Lotker, Z., Peleg, D.: Assortative mixing equilibria in social network games. CoRR, abs/1703.08776 (2017)
3. Avin, C., Keller, B., Lotker, Z., Mathieu, C., Peleg, D., Pignolet, Y.-A.: Homophily and the glass ceiling effect in social networks. In: Proceedings of the 2015 Conference on Innovations in Theoretical Computer Science, pp. 41–50. ACM (2015)
4. Barabási, A.-L., Albert, R.: Emergence of scaling in random networks. Science **286**(5439), 509–512 (1999)
5. Bilbao, J.M.: Cooperative Games on Combinatorial Structures, vol. 26. Springer, US (2012). doi:10.1007/978-1-4615-4393-0
6. Bramoulle, Y., Currarini, S., Jackson, M.O., Pin, P., Rogers, B.W.: Homophily and long-run integration in social networks. J. Econ. Theor. **147**(5), 1754–1786 (2012)
7. Chen, H.-L., Roughgarden, T.: Network design with weighted players. Theor. Comput. Syst. **45**(2), 302–324 (2009)
8. Currarini, S., Jackson, M.O., Pin, P.: An economic model of friendship: homophily, minorities, and segregation. Econometrica **77**(4), 1003–1045 (2009)
9. Di Stefano, A., Scatà, M., La Corte, A., Liò, P., Catania, E., Guardo, E., Pagano, S.: Quantifying the role of homophily in human cooperation using multiplex evolutionary game theory. PLoS ONE **10**(10), e0140646 (2015)
10. Fu, F., Nowak, M.A., Christakis, N.A., Fowler, J.H.: The evolution of homophily. Scientific reports, 2 (2012)
11. Jackson, M.O.: A survey of network formation models: stability and efficiency. In: Group Formation in Economics: Networks, Clubs, and Coalitions, pp. 11–49 (2005)

12. Jackson, M.O., et al.: Social and Economic Networks, vol. 3. Princeton University Press, Princeton (2008)
13. Jackson, M.O., López-Pintado, D.: Diffusion and contagion in networks with heterogeneous agents and homophily. Netw. Sci. **1**(01), 49–67 (2013)
14. Jackson, M.O., Watts, A.: The evolution of social and economic networks. J. Econ. Theor. **106**(2), 265–295 (2002)
15. Lazarsfeld, P.F., Merton, R.K., et al.: Friendship as a social process: a substantive and methodological analysis. Freedom Control Mod. Soc. **18**(1), 18–66 (1954)
16. McPherson, M., Smith-Lovin, L., Cook, J.M.: Birds of a feather: homophily in social networks. Annu. Rev. Sociol. **27**, 415–444 (2001)
17. Newman, M.E.J.: Mixing patterns in networks. Phys. Rev. E **67**, 026126 (2003)
18. Tardos, E., Wexler, T.: Network formation games and the potential function method. In: Algorithmic Game Theory, pp. 487–516 (2007)

Nash Equilibrium and Stability in Network Selection Games

Mohit Hota$^{(\boxtimes)}$ and Sanjiv Kapoor

Illinois Institute of Technology, Chicago, IL 60616, USA
mhota@hawk.iit.edu, kapoor@iit.edu

Abstract. We study network selection games in wireless networks. Each client selects a base station to maximize her throughput. We utilize a model which incorporates client priority weight and her physical rate on individual Base Stations. The network selection behavior considered is atomic, implying that a client connects to exactly one Base Station.

We formulate a non-cooperative game and study its convergence to a pure Nash equilibrium, if it exists, or prove non-existence otherwise, and present algorithms to discover pure Nash equilibrium for multiple cases.

1 Introduction

Enhancements in wireless connectivity involve the ability to choose the best available network connection. This is evident in recently put forth proposals and implementations, where a wireless device selects the provider (base station) and type of access (Wi-Fi, WiMax or GPRS schemes, femto etc.) which permits the best speed or rate, on the basis of location and availability (Google Fi services is an example). Moreover, priority weights, ensuring individual user priorities according to fixed agreements, are being increasingly suggested by providers. These priority agreements would serve to provide Quality of Service (QoS).

Throughput analysis of accessing heterogeneous radio technologies has been studied in [1–4] where clients utilize information from the access networks (termed RAT or RAN) to determine the choice of network access points (also referred to as Base stations). The standard approach is to consider the clients to be autonomous agents. Alternately, rules can be imposed on the RAN clients to regulate traffic.

The key decision for users in such a model is the selection of the network access point. The system of autonomous agents competing for a limited set of resources gives rise to a congestion game. Such a system leads to the formation of a complex system model where a user (client) would select, based on priority weights, a provider's base station and an instantaneous PHY rate provided by the base station, as has been utilized in [4], depending on current physical conditions like base station load, location or even radio bandwidth congestion. All such factors would determine the throughput that a client would be able to obtain on a base station.

Every client seeks to maximize her own total throughput without regard for how other clients are affected by her actions and thus, we formulate a game

© ICST Institute for Computer Sciences, Social Informatics and Telecommunications Engineering 2017
L. Duan et al. (Eds.): GameNets 2017, LNICST 212, pp. 40–50, 2017.
DOI: 10.1007/978-3-319-67540-4_4

where each client behaves selfishly to maximize her throughput. Such a game-theoretic model has previously been studied also in [3,4]. Additional throughput or utility models can be found in the survey paper [1]. We term the above model as an *atomic throughput game*, also termed as a RAN selection game in previous papers. These previous papers leave a number of unresolved issues regarding the existence of pure Nash equilibrium (interchangeably, for simplicity, referred to as *Nash equilibrium* in this paper) in the defined games.

The RAN selection game falls into the class of congestion games. Atomic congestion games with a cost function dependent on the number of clients occupying a resource were first studied in [5] with consequent work on client specific utilities in [6]. The computational complexity of determining Nash equilibrium in these games was studied in [7], where they showed that atomic congestion games with arbitrary cost functions are $\mathcal{PLS}$-Complete. Wireless congestion games and cost network specific cost functions have been studied in [2,8,9]. Unlike the prior studies, the model in [3,4] utilizes the throughput itself as a metric of performance. Additional game theoretic models using evolutionary games [10,11] have been studied but are not relevant as these models correspond to non-atomic versions of the game with large number of users, each with infinitesimal impact.

In this paper we consider the RAN selection game:

- We first show that pure Nash equilibrium does not always exist for the RAN selection game with non-uniform weights and rates, implying that the system might not stabilize at all. This resolves a question left unanswered in [3,4] where Aryafar et al. alluded to such a result. Resolving the existence and complexity of Nash equilibrium is considered important as it characterizes the convergence towards stability of such autonomous systems. We consider interesting practical cases and prove that pure Nash Equilibrium always exists if the user has uniform or identical priorities over all base stations. We provide an ϵ-approximate Nash Equilibrium algorithm which runs in polynomial time in this case, as well as a polynomial algorithm to compute pure Nash equilibrium when, additionally, rates are uniform.
- We consider *priority regulated games*, where priority can be used to regulate the throughput rate and disprove a conjecture from [4] which states that a Nash Equilibrium always exists in games where the priority weights is a polynomial function of the rates. On the positive side, we provide a simple fairness rule that ensures convergence to a solution which is stable, i.e. no further improvements are possible. This stable point may not be a Nash equilibrium of the original strategy space but the system is stable under the rule.

1.1   Network Model

The wireless selection problem has a set of clients P accessing a set of wireless access points, which we refer to as Base Stations, K. The base stations represent the range of wireless access points, Wi-Fi and GPRS etc. Each client accesses a base station and negotiates a rate of access. The wireless selection problem is that of scheduling clients to base stations to optimize throughput. We represent the wireless selection problem by a network model where the underlying graph is a bipartite graph represented by $G = (P, K, E)$. The clients are represented by one (independent) vertex set P and the set of Base Stations (BS), K, the second (independent) vertex set. The set of edges E represents the base stations available between the clients in the set P and the base stations. An edge $e = (i, k), i \in P, k \in K$ exists if and only if the client i can access base station k. Each client i is characterized by two parameters, the weight $\phi_{i,k}$ that provides her a priority on a base station $k \in K$ and the PHY rate $R_{i,k}$ that she can obtain on that base station k. The throughput that the clients acquire from the base station k is dependent on the other clients that utilize the base station.

Throughput Model. The throughput model we use is based on the model in [3, 4] that defines the throughput client i obtains on base station k as

$$\omega_{i,k} = \frac{\phi_{i,k}}{\sum_{j \in s(k)} \frac{\phi_{j,k}}{R_{j,k}}}$$

where $s(k)$ is the set of clients that are currently accessing base station k.

Since each client has an independent choice of scheduling her traffic on the available base stations, the rational autonomous decisions of the client can be modeled by a game:

A **Throughput Game** is denoted by $TG(P, K, \phi, R)$ where P are the clients (clients) in the game, K is the set of base stations, $\phi : P \times K \to \mathbb{R}^+$ is a function representing the priorities (weights) of clients on the base stations K, $R : P \times K \to \mathbb{R}^+$ is a function representing the rates the clients have obtained on the base stations. In a throughput game, a client selects one base station to transfer data, and given the selection of the other clients, selfishly selects the base station on which she receives maximum throughput. We also consider restricted models defined below:

1. Different types of traffic require a priority that is dictated by their type, e.g., video traffic requires a certain priority level, and do not depend on the base stations, leading to **Uniform Priority Throughput** games, denoted by $TG_P(P, K, \phi, R)$, where the priority levels are independent of the base stations, i.e. $\phi_{i,k} = \phi_{i,k'} = \phi_i, \forall k, k'$.
2. Furthermore, devices may only be able to communicate at a particular rate, leading to **Uniform Rate Throughput** games, denoted by $TG_R(P, K, \phi, R)$, where the rates achieved by a client i is independent of the base stations, i.e. $R_{i,k} = R_{i,k'} = R_i, \forall k, k'$.

We define the *Load* on a base station k, when a set $s(k)$ of clients are scheduled on base station k, to be $G_{s(k)} = \sum_{j \in s(k)} \frac{\phi_{j,k}}{R_{j,k}}$ where the contribution of a client j to the load is $\frac{\phi_{j,k}}{R_{j,k}}$. Note that when i is the only client on a base station k, she gets throughput $\omega_{i,k} = R_{i,k}$.

A *pure Nash equilibrium* is defined as an assignment of clients to base stations such that no client can unilaterally improve her throughput by switching to a different base station.

2   Nash Equilibrium in Throughput Games

We first resolve the question of existence of Nash equilibrium in throughput games, left unanswered in [4].

Theorem 1. *There exists a throughput game, $TG(P, K, \phi, R)$, for which there is no pure Nash equilibrium.*

Proof. In order to determine an example for a game, a Monte Carlo algorithm was used to generate the base station rates and priorities. In a game involving 3 clients and 3 base stations, the following values of ϕ and R present a scenario such that no configuration of assignments result in any client being satisfied on the base station she occupies. The matrix Φ represents the priorities $\phi_{i,k}$ and the matrix R represents the rates $R_{i,k}$.

$$\Phi = \begin{bmatrix} & L_1 & L_2 & L_3 \\ P_1 & 9.8 & 1.6 & 5.1 \\ P_2 & 8.1 & 0.2 & 8.6 \\ P_3 & 4.6 & 3.9 & 8.8 \end{bmatrix} \quad R = \begin{bmatrix} & L_1 & L_2 & L_3 \\ P_1 & 98.3 & 80.8 & 12.6 \\ P_2 & 27.6 & 32.6 & 21.2 \\ P_3 & 65.8 & 14.9 & 9.8 \end{bmatrix}$$

Any configuration in this instance results in cycling. To illustrate one such cycle, consider an initial configuration $(2, 1, 1)$ denoting that client 1 is on link 2, client 2 is on link 1 and client 3 is on link 1. Client 3 can obtain a higher throughput than what she already has by moving from link 1 to link 2, and does so. The configuration is thus $(2, 1, 2)$, following which, Client 1 then switches to link 1 to obtain a higher throughput, yielding the configuration $(1, 1, 2)$. The configuration keeps changing and eventually cycles back to a previous state. The cycle is shown in Fig. 1.

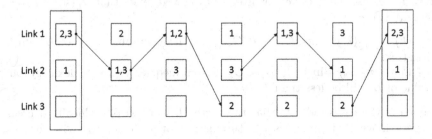

Fig. 1. Client cycling in a throughput game where Nash equilibrium does not exist

2.1   Nash Equilibrium in Uniform Priority Throughput Games

Since we have shown that a Nash equilibrium may not always exist in throughput games, we study its existence in Uniform Priority Throughput Games.

Theorem 2. *Every instance of a Uniform Priority Throughput game, $TG_U(P, K, \phi, R)$ (where $\phi_{ik} = \phi_{il} = \phi_i$), has a pure Nash equilibrium.*

Proof. Given an assignment of clients to base stations, characterized by specifying $s(k)$, the set of clients on base station k, we first establish an inequality, which provides a condition under which a client switches to another base station. Consider a client i who chooses to make a move from k to k' to get a higher throughput. For this move to occur, we must have

$$\frac{\phi_i}{\sum_{j \in s(k')} \frac{\phi_j}{R_{j,k'}} + \frac{\phi_i}{R_{i,k'}}} > \frac{\phi_i}{\sum_{j \in s(k)} \frac{\phi_j}{R_{j,k}}} \tag{1}$$

The load on base station k is $\sum_{j \in k} \frac{\phi_j}{R_{j,k}} = G_{s(k)}$, as defined in the network model. Inequality (1) then becomes

$$G_{s(k')} + \frac{\phi_i}{R_{i,k'}} < G_{s(k)} \text{ or equivalently, } G_{s'(k')} < G_{s(k)} \tag{2}$$

where $G_{s'(k')} = G_{s(k')} + \frac{\phi_i}{R_{i,k'}}$ is the load of base station k after i moves.

This expresses the fact that when client i moves from base station k to k' to increase her throughput, the load on base station k' after the move must be less than the pre-move load of k, otherwise client i would have had no incentive to move.

We then consider a vector $L = \{\{G_{s(k_1)}, \cdots, G_{s(k)}, G_{s(k')}, \cdots, G_{s(k_K)}\}$ s.t. $\{G_{s(k_1)} > \cdots > G_{s(k)} > G_{s(k')} > \cdots > G_{s(k_K)}\}\}$, i.e., k_1 is the base station with the highest load and k_K is the base station with the smallest load. Our claim is that the load vector L, which is the sorted loads of base stations, decreases (in lexicographic ordering) for every move that client i makes to increase her throughput. We prove our claim below:

We define the position (increasing from left to right) of the load of a base station k in a load vector L by $\pi_L(k)$. Let L be the load vector before client i moves and L' be the load vector after i has moved. Note that $\pi_L(k) \leq \pi_{L'}(k)$. There are two cases:

1. $\pi_{L'}(k) < \pi_{L'}(k')$: Since $G_{s'(k)} < G_{s(k)}$, the lexicographic value of L' will be less than L.
2. $\pi_{L'}(k') < \pi_{L'}(k)$: Since $G_{s'(k')} < G_{s(k)}$ from inequality (2), the lexicographic value of L' will be less than L.

Each of the cases indicate that the vector L will lexicographically decrease for every move that improves the throughput of a client. To show that the minimum load on each base station is lower bounded by a positive value, let $\phi_{\min} = \min_{i \in P} \phi_i$ and $R_{\max} = \max_{i \in P, k \in K} R_{i,k}$. The minimum load on each base station, which is occupied by at least one client, is then at least $\frac{\phi_{\min}}{R_{\max}}$. Therefore, the *uniform throughput priority game* will always converge to a pure Nash equilibrium.

2.2 ϵ-Approximate Nash Equilibrium for Uniform Priority Models

Based on the proof of Theorem 2, we observe that Nash equilibrium can be determined by allowing clients to improve their throughput by switching base stations. The number of improvement steps of the lexicographic ordering in vector L is upper bounded by $O\left(|P| \times \frac{\sum_{i \in P, k \in K} \frac{\phi_i}{R_{i,k}}}{\delta}\right)$, where δ is the minimum change in lexicographic value caused by a switch. When the values of ϕ_{ij} and R_{ij} are integers, $\delta \geq \frac{1}{R_{max}^2}$. Similar bounds can be established for rationals. Finding a polynomial time algorithm for determining Nash equilibrium appears difficult. Therefore, we specialize certain parameters to obtain faster algorithms for achieving a near-Nash equilibrium state.

We first define an ϵ-**approximate Nash equilibrium**: A throughput game is at an ϵ-approximate Nash equilibrium if for every client, a switch to another base station improves her throughput by a factor of at most $(1 + \epsilon)$.

Algorithm 1. Finding ϵ-approximate Nash equilibrium in $TG_P(P, K, \phi, R)$

1: Start with any random assignment of clients, where the set of clients on base station k is given by $s(k), \forall k$.
2: **while** $\exists$ clients who can improve their throughput by a factor of $(1+\epsilon)$ by switching to another base station **do**
3:　　Select client i s.t. $(i, k_i') = \arg \min_{(i,k'(i))} (G_{s'(k_i)} + G_{s'(k_i')} + \sum_{k \in K, k \neq k_i, k_i'} G_{s(k)})$
　　　 where i is assigned to k_i and moves to k_i', and $s'(k)$ denotes the set of clients on k after the movement of i.
4:　　Move i from k_i to k_i'.
5: **end while**

Theorem 3. *Given an instance of a Uniform Priority Throughput Game, Algorithm 1 finds a ϵ-approximate Nash equilibrium in time $O(t \times P \times K)$, where $t = \log_{1+\epsilon} \frac{R_{max}}{\phi_{min}} + \log_{1+\epsilon} \sum_{i \in P} (\frac{\phi}{R})_{i_{max}}$ is the upper bound on the number of steps a client moves, where $(\frac{\phi}{R})_{i_{max}} = \max_{k, k' \in K} \frac{\phi_k}{R_{k'}}$.*

Proof. We have already established that a switch implies a lexicographic decrease of vector L, as shown in Theorem 2, and therefore, assured that the approximate Nash equilibrium is achieved by the algorithm.

To calculate the time complexity, we provide an upper bound t on the number of times a client would have to switch to reach her final choice of base station. Since a client i can only move if she gains a factor of $(1 + \epsilon)$ on her current throughput, t can be calculated by comparing the lower bound and upper bound, termed $\omega_{i_{min}}$ and $\omega_{i_{max}}$, respectively, on her possible throughput.

We obtain the value of $\omega_{i_{max}}$ for a client i by placing her alone on the base station where she has the maximum PHY rate $R_{max} = \max_{i \in P, k \in K} R_{i,k}$, since the load on the base station increases as soon as she shares a base station with another client. Therefore, $\omega_{i_{max}} = R_{max}$. Similarly, we get $\omega_{i_{min}}$ by placing the client with the minimum ϕ_i (ϕ_{min}) with all the other clients in the game, and then by selecting the maximum load contribution of each client, $(\frac{\phi}{R})_{i_{max}} = \max_{k \in K} (\frac{\phi_i}{R_{i,k}})$, giving a total load of $\sum_{i \in P} (\frac{\phi}{R})_{i_{max}}$. Therefore,

$\omega_{i_{min}} = \frac{\phi_{min}}{\sum_{i \in P} (\frac{\phi}{R})_{i_{max}}}$. We then obtain the bound on t by using the fact that when the algorithm terminates, the maximum throughput is at most $\frac{\omega_{i_{max}}}{\omega_{i_{min}}}$. Therefore, $(1+\epsilon)^t \leq \frac{\omega_{i_{max}}}{\omega_{i_{min}}}$, which implies $t \leq \log_{1+\epsilon} \frac{R_{max}}{\phi_{min}} + \log_{1+\epsilon} \sum_{i \in P} (\frac{\phi}{R})_{i_{max}}$ which then leads to our result.

2.3   Finding Equilibrium in Uniform Priority-and-Rate Games

While a Nash equilibrium is not easily (in polynomial time) found in Uniform Priority games, we show that by altering the uniform priority game to include uniform rates, denoted by $TG_{P,R}(P, K, \phi, R)$, a Nash equilibrium can be discovered by a polynomial time algorithm.

Algorithm 2. Finding a Pure Nash equilibrium in $TG_{P,R}(P, K, \phi, R)$

1: Sort clients in non-increasing order of $\frac{\phi_i}{R_i}$
2: **for** Client $i = 1 \cdots |P|$ **do**
3:    $k_i = \arg\min_{k \in K}(G_{s(k)} + \frac{\phi_i}{R_i})$
4:    Assign client i to base station k_i.
5: **end for**

Theorem 4. *Given an instance of a Uniform Priority-and-Rate Throughput game $TG_{P,R}(P, K, \phi, R)$, Algorithm 2 correctly finds a pure Nash equilibrium in time $O(|P|(|K| + log|P|))$.*

Proof. The algorithm assigns a new client to a base station and ensures that the client gets maximum throughput, given the current system configuration. For our algorithm to be correct, an addition of a new client to the system should not induce any moves.

First, we use contradiction to show that after addition of a new client to a base station, other clients from that base station do not have an incentive to move to other links. Let $G_{s(k)}$ and $G_{s(k')}$ be the loads of base stations k and k' respectively before either client i or i' have been introduced. Suppose that on addition of the ith client to base station k, client i' wants to switch from using base station k to k', implying inequality (3),

$$\frac{\phi_{i'}}{G_{s(k)} + \frac{\phi_i}{R_i} + \frac{\phi_{i'}}{R_{i'}}} < \frac{\phi_{i'}}{G_{s(k')} + \frac{\phi_{i'}}{R'_i}} \Rightarrow G_{s(k')} < G_{s(k)} + \frac{\phi_i}{R_i} \quad (3)$$

and from the fact that client i was previously assigned to base station k, we have inequality (4)

$$\frac{\phi_i}{G_{s(k)} + \frac{\phi_i}{R_i} + \frac{\phi_{i'}}{R_{i'}}} > \frac{\phi_i}{G_{s(k')} + \frac{\phi_i}{R_i}} \Rightarrow G_{s(k')} > G_{s(k)} + \frac{\phi_{i'}}{R_{i'}} \quad (4)$$

From the sorting performed $\frac{\phi_i}{R_i}$ in step 1, we have

$$\phi_{i'}/R_{i'} > \phi_i/R_i \tag{5}$$

Using the above inequalities, we get the following contradiction

$$G_{s(k')} < G_{s(k)} + \frac{\phi_i}{R_i} < G_{s(k)} + \frac{\phi_{i'}}{R_{i'}} \text{ and } G_{s(k')} > G_{s(k)} + \frac{\phi_{i'}}{R_{i'}} > G_{s(k)} + \frac{\phi_i}{R_i}$$

In the second case, we show that when a client p is added to a base station k, clients p' from other base stations k'(say) do not move to k. Prior to adding p, p' had chosen base station k' over base station k.

$$\therefore G_{s(k)} + \frac{\phi_{p'}}{R_{p'}} > G_{s(k')} + \frac{\phi_{p'}}{R_{p'}} \Rightarrow G_{s(k)} > G_{s(k')} \tag{6}$$

If it were beneficial for p' to switch to base station k now, the following inequality must be true

$$G_{s(k)} + \frac{\phi_p}{R_p} + \frac{\phi_{p'}}{R_{p'}} < G_{s(k')} + \frac{\phi_{p'}}{R_{p'}} \Rightarrow G_{s(k)} < G_{s(k')} \tag{7}$$

which is contradictory to inequality (6), thus proving our claim.

The running time of step 1 is $O(|P|\log|P|)$ for sorting the values. Step 2 has $|P|$ iterations of steps 3 and 4, which perform $|K|$ comparisons, thus giving a total running time of $O(|P|(\log|P| + |K|))$.

3   Nash Equilibrium in Rate-Dependent Priority Throughput Games

Throughput games where the priorities are a function of the rates have been investigated in [4]. In this model, $\phi_{i,k} = R_{i,k}^{\beta}$, and the game is denoted by $TG_\beta(P, K, \phi, R)$. Properties of this model have been studied in [4] where it was conjectured that Nash equilibrium exists for any value of β. We first disprove this conjecture and then provide a set of rules under which we prove that a stable point exists.

Theorem 5. *There exists an instance of a Rate-dependent Throughput Game, $TG_\beta(P, K, \phi, R)$, with $\phi_{ij} = R_{ij}^{-1.5}$ for which a pure Nash equilibrium does not exist.*

Proof. Similar to Theorem 1, we use a Monte Carlo algorithm to obtain the values of ϕ_{ij} and R_{ij} where $\phi_{ij} = R_{ij}^\beta$ and $\beta = -1.5$. The matrix R is

$$R = \begin{bmatrix} & L_1 & L_2 & L_3 \\ P_1 & 8.719 & 3.755 & 4.927 \\ P_2 & 5.802 & 1.361 & 5.783 \\ P_3 & 4.824 & 1.094 & 4.643 \\ P_4 & 3.340 & 9.648 & 8.743 \\ P_5 & 2.818 & 9.543 & 4.325 \end{bmatrix}$$

Values of ϕ can be generated using matrix R and $\beta = -1.5$. In fact, such examples were found for multiple values of β where $\beta < 0$. We illustrate an instance of a cycle of configurations in Fig. 2. Each configuration of the above instance results in similar cycles, yielding a system where no pure Nash equilibrium exists.

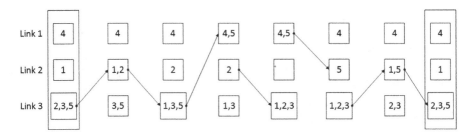

Fig. 2. Client cycling in a rate dependent throughput game where Nash equilibrium does not exist

3.1   Conditions for Convergence to a Stable Point

Having established that Nash equilibrium need not always exist, we now establish a protocol that ensures convergence to a stable point, thus preventing thrashing in the system.

Fair-Movement Protocol:

The following rules shall apply:

1. When a new client joins a system, she will be automatically assigned to the base station she has the highest rate on.
2. For every client, say i, switching from base station k to k' is permitted, only if $R_{i,k'} \leq R_{i,k}$. This is termed as the *Fair Movement Rule*

The purpose to the *Fair Movement Rule* is that since a client has already sought to reject a base station she was assigned a higher rate on, she must not be allowed to act selfishly with respect to her base rate $R_{i,k}$ and prevent the system from stabilizing.

Theorem 6. *Under the* **Fair-Movement Rule***, every Rate-dependent Throughput Game has a stable point; i.e., no client gains by unilaterally changing to a different assignment.*

Proof. Consider a vector L of loads on base stations $L = \{Gs_{(k_1)}, Gs_{(k_2)}, \cdots, Gs_{(k_K)}\}$ s.t. $\{G_{s(k_1)} > ... > G_{s(k)} > G_{s(k')} > ... > G_{s(k_K)}\}\}$. Note that the load of a base station is now given by $G_{s(k)} = \sum \dfrac{1}{R_{i,k}^{1-\beta}}$

Now, client i moves from base station k to k' ($G_{s(k)}$ is inclusive of client i) when she gets a higher throughput on k',

$$\frac{Gs_k}{R_{i,k}^{\beta}} > \frac{1}{R_{i,k'}^{\beta}}(Gs_{k'} + \frac{1}{R_{i,k'}^{1-\beta}}) \tag{8}$$

Using rule 2 of the Fair-Movement Protocol, we have $R_{i,k}^{\beta} - R_{i,k'}^{\beta} > 0$, therefore

$$\frac{1}{R_{i,k'}^{\beta}}(Gs_{k'} + \frac{1}{R_{i,k'}^{1-\beta}}) > \frac{1}{R_{i,k}^{\beta}}(Gs_{k'} + \frac{1}{R_{i,k'}^{1-\beta}}) \tag{9}$$

So, from (8) and (9), $\frac{1}{R_{i,k}^{\beta}}Gs_k > \frac{1}{R_{i,k}^{\beta}}(Gs_{k'} + \frac{1}{R_{i,k'}^{\beta}}) \Rightarrow Gs_k > Gs_{k'} + \frac{1}{R_{i,k'}^{1-\beta}}$, implying that the vector L decreases in lexicographic ordering every time a client i switches from base station k to k'. Thus, the game will be stable.

4   Conclusions and Acknowledgements

This paper has presented results for pure Nash equilibrium in a wireless game model where throughput has been used as the measure of the payoff. It would be of further interest to include link access costs, client budgets and general utility functions in the model.

The research was supported in part by NSF grant: CCF-1451574.

References

1. Kuo, G.S.: Mathematical modeling for network selection in heterogeneous wireless networks - a tutorial. IEEE Commun. Surv. Tutorials **15**(1), 271–292 (2013)
2. Malanchini, I., Cesana, M., Gatti, N.: Network selection and resource allocation games for wireless access networks. IEEE Trans. Mob. Comput. **12**(12), 2427–2440 (2013)
3. Aryafar, E., Keshavarz-Haddad, A., Wang, M., Chiang, M.: Rat selection games in hetnets. In: INFOCOM, pp. 998–1006. IEEE (2013)
4. Monsef, E., Keshavarz-Haddad, A., Aryafar, E., Saniie, J., Chiang, M.: Convergence properties of general network selection games. In: INFOCOM, pp. 1445–1453. IEEE (2015)
5. Rosenthal, R.W.: A class of games possessing pure-strategy nash equilibria. Int. J. Game Theory **2**, 65–67 (1973)
6. Milchtaich, I.: Weighted congestion games with separable preferences. Games Econ. Behav. **67**(2), 750–757 (2009)
7. Fabrikant, A., Papadimitriou, C., Talwar, K.: The complexity of pure nash equilibria. In: Proceedings of the Thirty-Sixth Annual ACM Symposium on Theory of Computing, pp. 604–612. ACM (2004)
8. Tekin, C., Liu, M., Southwell, R., Huang, J., Ahmad, S.H.A.: Atomic congestion games on graphs and their applications in networking. IEEE/ACM Trans. Netw. **20**(5), 1541–1552 (2012)

9. Ibrahim, M., Khawam, K., Tohm, S.: Congestion games for distributed radio access selection in broadband networks. In: GLOBECOM, pp. 1–5. IEEE (2010)

10. Shakkottai, S., Altman, E., Kumar, A.: Multihoming of users to access points in wlans: a population game perspective. IEEE J. Sel. Areas Commun. **25**(6), 1207–1215 (2007)

11. Niyato, D., Hossain, E.: Dynamics of network selection in heterogeneous wireless networks: an evolutionary game approach. IEEE Trans. Veh. Technol. **58**(4), 2008–2017 (2009)

On the Finite Population Evolutionary Stable Strategy Equilibrium for Perfect Information Extensive Form Games

Aycan Vargün[(✉)] and Mehmet Emin Dalkılıç[(✉)]

Ege University, Bornova, 35100 İzmir, Turkey
91130000040@ogrenci.ege.edu.tr, mehmet.emin.dalkilic@ege.edu.tr
http://www.ube.ege.edu.tr/~dalkilic

Abstract. This study presents an adaptation of finite population evolutionary stable strategy definition by Schaffer in [1,2] to perfect information extensive form games. In this adaptation, players reach a finite population evolutionary stable strategy equilibrium by using finite population evolutionary stable strategies which ensure that the game ends up with equal payoffs. We studied the fpESS equilibria of some famous two-player bargaining games such as the ultimatum game, the dictatorship game and a dollar auction game. Not all Perfect Information Extensive form games have an fpESS equilibrium. However, when there exist an fpESS equilibrium in these games, the outcome is a perfectly fair one; that is, all players get equal payoffs.

Keywords: Perfect information extensive form game · Ultimatum game · Fairness · Finite population evolutionary stable strategy

1 Introduction

Perfect Information Extensive Form games are very important in game theory. As one of them, the ultimatum game is a widely researched problem. There is some amount of money that the first player is asked to divide between himself and the second player. If the second player does not accept his share, he rejects it and both players take nothing. If he accepts the offer, both players take the amounts that they hold. The dictatorship game is also widely researched in game theory. The first player determines the shares again, but the second player cannot reject the offer.

When we look at the experiments on the ultimatum game, the results that we encounter are very different than the theory predicts. Second players often reject the offers less than half of the money and first players are willing to offer much more than the least that they can offer [3,4]. However, when we look at the experimental results of the dictatorship game, first players are more selfish and they offer much less compared to the offers in the ultimatum game.

The first and the second player consider each other's actions in the perfect information extensive form games. Even if it does not seem rational, it is important to get greater or equal payoff for players. Evolutionary Stable Strategies

© ICST Institute for Computer Sciences, Social Informatics and Telecommunications Engineering 2017
L. Duan et al. (Eds.): GameNets 2017, LNICST 212, pp. 51–59, 2017.
DOI: 10.1007/978-3-319-67540-4_5

(ESS) take this point into account. Therefore, we decided to study fpESS (finite population ESS) approach to analyze the perfect information extensive form games and especially the ultimatum and the dictatorship games. We show that for some perfect information extensive form games, there are some finite population evolutionary stable strategies that a player can quarantee a payoff at least as large as any opponent's payoff. When players pick one of these strategies, they can prevent to be beaten by their opponents (receiving a smaller payoff than any of the opponent's).

In the rest of the paper we first give, in Sect. 2, the relevant background and definitions for the application of fpESS to Extensive Form games followed by, in Sects. 2.1 and 2.2, the work to find the fpESS equilibria for three well known game instances: the ultimatum game, the dictatorship game and the dollar auction game.

2   Adaptation of FpESS to Extensive Form Games in Induced Form

The fpESS concept is introduced in [5] and restated as the following definition.

Definition 1. *Let S be the strategy set of a symmetric normal form game. An fpESS s is a strategy in which $\forall s^* \in S, u(s, s^*) \geq u(s^*, s)$ (by Schaffer [1, 2]).*

In the fpESS concept, the game is symmetric and the players have the same strategy sets. We applied this approach to the induced normal forms of extensive form games. It is obvious that an induced normal form does not have to be symmetric. However, we can apply this definition to the induced normal form of extensive form games.

Definition 2. *Let S_1, S_2, ..., S_n be the strategy sets in the induced normal form of a perfect information extensive form game with n players. A strategy $s_i \in S_i$ is an fpESS if $\forall j, u_i(s_i, s_{-i}) \geq u_j(s_j, s_{-j})$ for all $s_{-i} \in S_{-i}$ and $s_{-j} \in S_{-j}$ where $S_{-i} = (S_1 \times ...S_{i-1} \times S_{i+1}... \times S_n)$ and $S_{-j} = (S_1 \times ...S_{j-1} \times S_{j+1}... \times S_n)$.*

Definition 2 implies that a strategy s_i is an fpESS for player i if there is no strategy available to any opponent that returns a greater payoff than that of the ith player's payoff.

Definition 3. *If $\forall i$ s_i^* is an fpESS, then $(s_1^*,..., s_n^*)$ is an fpESS equilibrium.*

Example 1. Consider a two player game with the payoffs $z_1 = (10, 10), z_2 = (30, -30), z_3 = (-20, 20)$. This extensive form game tree is shown in Fig. 1(a). When the induced normal form of this game is obtained as shown in Table 1, we see from the row labeled s_1, i.e., $[(10, 10), (30, -30)]$, that the first players payoff is always greater than or equal to the second player. In other words strategy s_1 is an fpESS for the first player. Similarly from the column s_3, which is $[(10, 10), (-20, 20)]$, under this strategy the second player's payoff is at least as big as its opponent, and thus s_3 is an fpESS for the second player. Thus, (s_1, s_3) is an fpESS equilibrium.

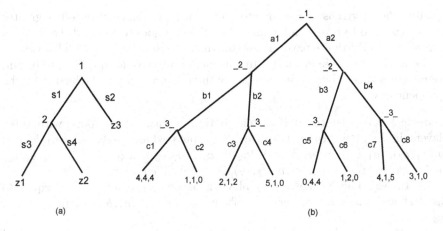

Fig. 1. Two examples of perfect information extensive form games (a) A two player game, (b) A three player game

Table 1. Induced normal form of the game in Fig. 1(a) with the payoffs $z_1 = (10, 10), z_2 = (30, -30), z_3 = (-20, 20)$

	s_3	s_4
s_1	$(10, 10)$	$(30, -30)$
s_2	$(-20, 20)$	$(-20, 20)$

In a perfect information extensive form game, there does not have to be an fpESS equilibrium.

Example 2. Assume that in Fig. 1(a), $z_1 = (10, -10), z_2 = (-20, -30), z_3 = (0, 0)$. Although the first player has two fpESSs (s_1 and s_2), the second player has no fpESS. Therefore, there is no fpESS equilibrium for this game. This shows that a player may not have an fpESS and a game may not have an fpESS equilibrium.

Example 3. We can analyze some famous cases. Pyrrhic victory is one of them in which the result is so devastating that the victor loses everything except the victory. This devastating lost is equal to defeat. We applied this result to the game tree in Fig. 1(a). The terminal nodes are $z_1 = (a_1, b_1), z_2 = (a_2, b_2), z_3 = (a_3, b_3)$.

Let S_1 be going to war, S_2 be not going to war as the strategies of the king Pyrrhus of Epirus and S_3 be accepting the challenge and going to war against Pyrrhus, S_4 be surrendering without battling as the strategies of the Romans at Heraclea in 280 BC.

In this game, if we assume that $a_1 = b_1$, $a_2 > b_2, a_3 \leq b_3$, we see that *(going to war as Pyrrhus, going to war as the Romans)* is an fpESS equilibrium. Here our assumption is that when both sides go to war, they will lose everything equally, when Pyrrhus choose not to go to war, the Romans' payoff is greater than or equal to Pyrrhus, and finally when Pyrrhus goes to war and the Romans surrender, the Pyrrhus's payoff is greater than the Romans' payoff. This victory can be modeled in different ways. In this model, we have an fpESS equilibrium. We interpret this as both sides can do anything in order not to be beaten by the opponent in war.

Example 4. Consider the perfect information extensive form game with three players in Fig. 1(b). Here, there are 8 outputs whose paths are (a_1, b_1, c_1), (a_1, b_1, c_2), (a_1, b_2, c_3), (a_1, b_2, c_4), (a_2, b_3, c_5), (a_2, b_3, c_6), (a_2, b_4, c_7), (a_2, b_4, c_8) in this game. $[a_1, (b_1, b_3), (c_1, c_3, c_5, c_7)]$ is an fpESS equilibrium.

In Tables 2 and 3, the second player's payoff is greater than or equal to the first and the third players' payoffs in (a_1, b_1, c_1), (a_1, b_1, c_2), (a_2, b_3, c_5) , (a_2, b_3, c_6).

The third player's payoff is greater than or equal to the first and the second player' payoffs in (a_1, b_1, c_1), (a_1, b_2, c_3), (a_2, b_3, c_5), (a_2, b_4, c_7).

The first player's payoff is greater than or equal to the second and third players' payoffs in (a_1, b_1, c_1), (a_1, b_1, c_2), (a_1, b_2, c_3), (a_1, b_2, c_4).

It may be hard to find a game with an fpESS equilibrium in which all players participate. However, if some of the players have fpESS strategies, they can choose them to play mutually.

Table 2. When the first player selects a_1

	(c_1, c_3)	(c_1, c_4)	(c_2, c_3)	(c_2, c_4)
b_1	(a_1, b_1, c_1)	(a_1, b_1, c_1)	(a_1, b_1, c_2)	(a_1, b_1, c_2)
b_2	(a_1, b_2, c_3)	(a_1, b_2, c_4)	(a_1, b_2, c_3)	(a_1, b_2, c_4)

Table 3. When the first player selects a_2

	(c_5, c_7)	(c_5, c_8)	(c_6, c_7)	(c_6, c_8)
b_3	(a_2, b_3, c_5)	(a_2, b_3, c_5)	(a_2, b_3, c_6)	(a_2, b_3, c_6)
b_4	(a_2, b_4, c_7)	(a_2, b_4, c_8)	(a_2, b_4, c_7)	(a_2, b_4, c_8)

2.1   FpESS Equilibria of Ultimatum and Dictatorship Games

Ultimatum game is a widely researched bargaining problem. We used [6] to express the equilibria in the ultimatum game. In this game, one of the players must divide A dollar as $(A - x, x)$. He takes $A - x$ to himself, gives x to the other player. If the second player accepts this sharing, he takes x and the first player takes $A - x$. If he does not accept the sharing, all the players take *zero* as their payoffs.

Payoffs do not have to be integers. For any value of x, there is a subgame for the second player. In this case, we can analyze the second player's action for each x value. x can be *zero* or greater than *zero*. When x is *zero*, to say *yes* or *no* is indifferent for the second player. When x is greater than *zero*, the second player says *yes* because $x > 0$. By this, we have two optimal strategies for the second player. The first is to say *yes* for $x \geq 0$. The second is to answer *yes* for $x > 0$ and *no* for $x = 0$.

Assume that offered payoffs are not integers. For the second player's first optimal strategy, the first player must offer *zero*. For the second player's second optimal strategy, the first player must offer any value greater than *zero*. Thus, first player's optimal strategy (considering the optimal strategies for player 1) is to offer the smallest $x > 0$. However, if the offers are made in real numbers, there is no such smallest $x > 0$. However, if offered values are integer, for Example 1 cent as the least value, the first player must offer 1 cent to the second player.

In this game, the subgame perfect equilibrium is that the first player offers *zero* and the second player accepts this. When the offered values are integers, we have one more subgame perfect equilibrium so that the first player offers the least value $x > 0$ to the second player, the second player accept this. However, when we look at the experiments, we do not see the theoretical predictions (that is, the first player offers the least positive amount and the second player accepts this minimum offer) are realized. Instead, we encounter more fair outcomes where offers considerably higher than minimum are typical and minimum offers usually rejected.

When we investigate the ultimatum game for any amount N, we see the remarkable feature of fpESS's is that a player does not offer more than half in the first position and a player does not accept less than half. When two players pick fpESS strategies to play, game ends fairly.

Proposition 1. *In an ultimatum game, let $N \in \mathbf{R}$ be the total payoff to share. $\forall N \in \mathbf{R}$ a strategy $s \in S_1$ in the induced normal form is an fpESS if and only if it includes always to offer less than or equal to $N/2$ as first player's strategy.*

Proof. If $\forall N \in \mathbf{R}$ a strategy $s \in S_1$ in the induced normal form is an fpESS, then it includes always to offer less than or equal to $N/2$ as first player's strategy. Assume that s is an fpESS, but it does not include to offer less than or equal to $N/2$. There exists a cell in the induced matrix row in which the first player offers more than half. The first players payoff becomes less than the second players payoff. This is in contrast with the definition of fpESS. Our assumption is invalid.

If $\forall N \in \mathbf{R}$ a strategy $s \in S_1$ in the induced normal form includes always to offer less than or equal to $N/2$ as first player's strategy, then it is an fpESS. Assume that s includes always to offer less than or equal to $N/2$, but it is not an fpESS. There is a cell where s resides in which the first player's payoff is less than the second player's payoff. This is a contradiction; our assumption is invalid. $\square$

Proposition 2. *In an ultimatum game, $\forall N \in \mathbf{R}$ a strategy $s \in S_2$ in the induced normal form is an fpESS if and only if it includes always not to accept any offer less than $N/2$ as second player's strategy.* $\square$

Proof. If $\forall N \in \mathbf{R}$ a strategy $s \in S_2$ in the induced normal form is an fpESS, then it includes always not to accept any offer less than $N/2$ as second player's strategy. Assume that s is an fpESS, but it includes to accept an offer less than $N/2$ as second player's strategy. There exists a cell in the induced matrix column in which the second player accepts less than half. The second players payoff becomes less than the first players payoff. This is contrast with the definition of fpESS. Our assumption is invalid.

If $\forall N \in \mathbf{R}$ a strategy $s \in S_2$ in the induced normal form includes always not to accept any offer less than $N/2$ as second player's strategy, then it is an fpESS. Assume that s includes always not to accept any offer less than $N/2$, but it is not an fpESS. There is a cell where s resides in which the second player's payoff is less than the first player's payoff. However, we accept that s includes always not to accept any offer less than $N/2$ as second player's strategy. This is contradiction. Our assumption is invalid. $\qquad\square$

Proposition 3. *In an ultimatum game,* $\forall N \in \mathbf{R}$, *an fpESS equilibrium is an outcome in which* $s_1 \in S_1$ *and* $s_2 \in S_2$ *are fpESSs. The payoffs in the fpESS equilibrium are equal.* $\qquad\square$

Proof. If s_1 and s_2 are fpESSs, then there are two possible solutions for the game. If the first player offers $N/2$ to the second player, he accepts and the game ends $(N/2, N/2)$ which the payoffs which are equal. If the first player offers less than $N/2$ to the second player, he does not accept and the game ends with the payoffs with equal payoffs $(0, 0)$. $\qquad\square$

fpESS equilibrium finalizes the game so that players can't gain advantage over each other. However, when we cannot do anything in a game in order not to be beaten by our opponent, we do not have an fpESS and there is not an fpESS equilibrium in the game. An example of this type of game is dictatorship game where the second player has no power to affect the outcome of the game. Dictatorship game does not include an fpESS for the second player and does not have an fpESS equilibrium.

2.2 FPESS Equilibrium of an Instance of the Dollar Auction Game

The Dollar auction game is a sequential game designed by Martin Shubik [7] to show that players are led to make irrational decisions in a perfect information game. In the game, the winner and the second highest bidder pay the last dollar amount that they bid. The game starts with a randomly selected player. When the first player says 5 cents, the second player can escalade the number by saying 10 cents or he can give up the game. If he says 10 cents, the first player can escalade the number by saying 15 cents, or he can also give up the game. If the first player gives up the game at this point, he pays 5 cents and he wins nothing. The second player pays 10 cents and he wins a dollar.

However, one bids 1.05$, another may bid 1.10$. They escalade the number above 1$ in order not to risk losing the game because they must pay the last dollar amount they bid even if they lose. At this point, bidding above 1$ is not rational to win a dollar as prize.

An example of this game is presented in [8]. There is 3$ as the prize and the maximum amount that a player can bid is 4$. The original game does not have an upper limit, but here there is.

Example 5. We adapted above game so that 2$ is the prize and the maximum that a player can bid is 3$. When a player says 3$, he wins the game but loses 1$. The game tree is given in Fig. 2.

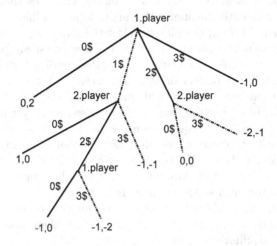

Fig. 2. A dollar auction game example with 2$ prize

We benefited from [9] to analyze the subgame perfect equilibrium for this game. The only difference between the game in Fig. 2 and the game in [9] is one of the first moves of first player. In our game, when the first player bids 0$, the game ends with the payoffs $(0, 2)$. On the other hand, in the game given in [9] if the first player bids 0$, the game ends with the payoffs $(0,0)$. The subgame perfect equilibria are $[(1\$, 0\$), (0\$, 0\$)]$, $[(2\$, 0\$), (2\$, 0\$)]$, $[(0\$, 0\$), (2\$, 0\$)]$, $[(1\$, 3\$), (0\$, 0\$)]$. These two games have the same subgame perfect equilibria because none of the first player's first moves in our game changes.

For example, the strategy that the first player will play in $[(1\$, 3\$), (0\$, 0\$)]$ is to play 1$ for the first move and to play 3$ for the last move in the tree. The strategy that the second player will play is to move 0$ for the subtree tied to 1$, to move 0$ for the subtree tied to 2$.

In this game, the first player has one fpESS. At the beginning, his move is 1$. If the second player plays 2$, he plays 3$. (1$, 3$) is the first player's fpESS. The second player has two fpESS's. These are (3$, 0$) and (3$, 3$). If the first player plays 1$, the second player plays 3$. If the first player plays 2$, the second player plays 0$ or 3$. The game ends up with $(-1, -1)$. We can interpret the second player's behavior as risking himself and the other player in order not to be beaten in the game. He does not want to gain less than or lose more than the opponent.

3 Conclusion

When we determine a strategy to play in a perfect information extensive form game, we consider our opponents' possible strategies. If we do not want to be beaten by our opponent at any cost, we use finite population evolutionary stable strategies if exist. It is hard to say that a player who uses these strategies is rational, but we frequently encounter this attitude in real life.

In an fpESS equilibrium, the players select the strategies so that they don't get less than their opponent. It guarantees that when we use an fpESS, the game will end up with a tie. This may bring a new understanding for player attitudes and their positions in the perfect information extensive form games.

In ultimatum game, the theoretical solutions do not explain experimental results thoroughly. In ultimatum games, when the second player picks a strategy that brings him to an fpESS equilibrium, he can't be worse off than the first player. When the first player knows this, he does not offer any unacceptable amount to the second player. He knows that any unfair (less than half the total amount) offer will be rejected. Any fpESS strategy which brings the players to an fpESS equilibrium ensures the game ends up fairly.

In dictatorship game, there is nothing to do for the second player when he does not want to gain less than the first player, thus there does not exist any fpESS and fpESS equilibrium in the dictatorship game.

In the instance of dollar auction game that we have analyzed, the fpESS equilibrium can be interpreted as any bidder can risk losing more and more by continuing with higher bids. A player can prefer staying in the game and losing equally (with the opponent) to withdrawing from the game earlier with relatively more loss.

We suppose there may be an intersection between subgame perfect equilibrium solution concept and fpESS equilibrium, it is a new research topic.

References

1. Schaffer, M.E.: Are profit-maximizers the best survivors? J. Econ. Behav. Organ. **12**, 29–45 (1989)
2. Schaffer, M.E.: Evolutionary stable strategies for a finite population and a variable contest size. J. Theor. Biol. **132**, 469–478 (1988)
3. Forsythe, R., Horowitz, J.L., Savin, N.E., Sefton, M.: Fairness in simple bargaining experiments. Games Econ. Behav. **6**(3), 347–369 (1991)
4. Oosterbeek, H., Sloof, R., Kuilen, G.: Cultural differences in ultimatum game experiments: evidence from a meta-analysis. Exp. Econ. **7**(2), 171–188 (2004)
5. Duersch, P., Oechssler, J., Schipper, B.C.: Pure Saddle Points and Symmetric Relative Payoff Games. Working Papers 0500, University of Heidelberg Department of Economics (2010)

6. Osborne, M.J.: An Introduction to Game Theory. Oxford University Press, New York (2003)
7. Shubik, M.: The dollar auction game: a paradox in noncooperative behavior and escalation. J. Conflict Resolut. **15**(1), 109–111 (1971)
8. Leininger, W.: Escalation and cooperation in conflict situations the dollar auction revisited. J. Conflict Resolut. **33**(2), 231–254 (1989)
9. Slantchev B.L.: http://slantchev.ucsd.edu/courses/gt/05-extensive-form.pdf

Application of Network Games

Designing Cyber Insurance Policies: Mitigating Moral Hazard Through Security Pre-Screening

Mohammad Mahdi Khalili[✉], Parinaz Naghizadeh, and Mingyan Liu

Electrical and Computer Science Department,
University of Michigan, Ann Arbor, USA
{khalili,naghizad,mingyan}@umich.edu

Abstract. Cyber-insurance has been studied as both a method for risk-transfer, as well as a potential incentive mechanism for improving the state of cyber-security. However, in the absence of regulated insurance markets or compulsory insurance, the introduction of insurance deteriorates network security. This is because by transferring part of their risk to the insurer, the insured agents can decrease their levels of effort. In this paper, we consider the design of insurance contracts by an (unregulated) profit-maximizing insurer, and allow for voluntary participation. We propose the use of pre-screening to offer premium discounts to higher effort agents. We show that such premium discrimination not only helps the insurer attain higher profits, but also leads the agents to improve their efforts. We show that with interdependent agents, the incentivized improvement in efforts can compensate for the effort reduction resulting from risk transfer, thus improving the state of network security over the no-insurance scenario. In other words, the availability of pre-screening signals benefits both the insurer, as well as the state of network security, without the need to regulate the market or compulsory participation.

1 Introduction

Organizations and businesses big and small are facing increasingly more complex, costly and frequent cyber threats. Many technology based protection methods such as novel cryptography schemes and protection softwares have been developed to reduce the risk of cyber threats. In addition to a myriad of technology based protection methods, cyber-insurance has emerged as an accepted risk mitigation mechanism, that allows purchasers of insurance policies/contracts to transfer their residual risks to the insurer.

The impact of cyber insurance on firms' security investment has been quite extensively studied in the past few years. These studies include cyber-insurance as a method for risk transfer, as well as a possible incentive mechanism for risk reduction, see e.g., [1–8]. Many papers on cyber insurance markets have studied the impact of cyber-insurance on the state of network security. Existing literature has arrived at two seemingly contradictory conclusions about the potential

---

This work is partially supported by the NSF under grant CNS-1616575.

© ICST Institute for Computer Sciences, Social Informatics and Telecommunications Engineering 2017
L. Duan et al. (Eds.): GameNets 2017, LNICST 212, pp. 63–73, 2017.
DOI: 10.1007/978-3-319-67540-4_6

of cyber-insurance as an incentive mechanism for risk reduction. The difference is mainly due to the underlying model of the insurer/insurance market. In particular, when the cyber-insurance market is modeled as a competitive market, e.g., [7,8], the insurance contracts are designed with the intention of attracting clients, and are hence not optimized to induce better security behavior. As a result, [7,8] show that the introduction of cyber-insurance deteriorates network security. Furthermore, as a consequence of the assumption of competitive markets, the insurers make no profit.

On the other hand, by considering a monopolist (profit-neutral) cyber-insurer, whose goal is to increase social welfare, [3–7] show that it is possible to design cyber-insurance contracts that lead users to improve their efforts toward securing their systems, and consequently, improve the state of security. The works in [5–7] propose *premium discrimination*; the idea is to assign less favorable contracts (i.e., higher premiums) to agents with worse types or lower efforts. These contracts can lead to an increase in social welfare and network security, as well as non-negative profit for the insurer. However, the underlying models assume that the insurer acts to increase social welfare (due to e.g., government regulation), and is therefore not profit-maximizing. In addition, participation by agents is assumed compulsory.

In this paper, we are similarly interested in the possibility of using cyber-insurance as an incentive mechanism for improved network security. We modify two of the key existing assumptions, in order to better capture the current state of cyber-insurance markets, by (1) considering a profit-maximizing cyber-insurer, and (2) ensuring that participation is voluntary, i.e., agents may opt out of purchasing a contract.

We propose the use of *pre-screening* (initial audit) by the insurer; pre-screening allows the insurer to evaluate the potential client's security posture, prior to offering the contract. This essentially allows the insurer to premium-discriminate the agents, based on their perceived/measured state of security. We provide sufficient conditions under which the introduction of pre-screening can lead to higher profits for the insurer, and that it also positively impacts the state of security. In other words, this type of pre-screening is a potential option for making cyber-insurance contracts better drivers for improved cyber-security.

2   A Single Risk-Averse Agent

We first consider a single-period contract design problem between a risk-neutral cyber-insurer and a risk-averse agent.[1] The agent exerts *effort* $e \in [0, +\infty)$ towards securing his system, incurring a cost of c per unit of effort. Let L_e denote the loss, a random variable, that the agent experiences given his effort e. We assume L_e has a normal distribution, with mean $\mu(e) \geq 0$ and variance $\lambda(e) \geq 0$. We assume $\mu(e)$ and $\lambda(e)$ are strictly convex, strictly decreasing, and twice differentiable. The decreasing assumption entails that increased effort

---

[1] Throughout the paper, we use she/her and he/his to refer to the insurer and agent(s), respectively.

reduces the expected loss, as well as its unpredictability, for the agent. The convexity assumption suggests that while initial investment in security leads to considerable reduction in loss, the marginal benefit decreases as effort increases. We assume once a loss L_e is realized, it will be observed by both the cyber-insurer and agent through e.g., reporting and auditing. We further assume $\lambda(e)$ is small compared to $\mu(e)$, so that $Pr(L_e < 0)$ is negligible. Finally, when the agent exerts an effort e, the insurer observes a *pre-screening* signal $S_e = e + W$, where W is a zero mean Gaussian noise with variance σ^2. This signal can be attained through, e.g., external audits or initial surveys filled out by the agent. We assume S_e is conditionally independent of L_e, given e.

Linear Contract and Insurer's Payoff: In this paper, we consider the design of a set of *linear* contracts. Specifically, the contract offered by the insurer consists of a base premium p, a discount factor α, and a coverage factor β. The agent pays a premium $p - \alpha \cdot S_e$, and receives $\beta \cdot L_e$ as coverage. We let $0 \le \beta \le 1$, i.e., coverage never exceeds the actual loss. Thus the insurer's utility (profit) is given by:

$$V(p, \alpha, \beta, e) = p - \alpha \cdot S_e - \beta \cdot L_e.$$

The insurer's expected profit is then given by $\overline{V}(p, \alpha, \beta, e) = p - \alpha e - \beta \mu(e)$.

Agent's Payoff without a Contract: If the agent chooses not to enter a contract, he bears the full cost of his effort as well as any loss. We assume

$$U(e) = -\exp\{-\gamma \cdot (-L_e - ce)\}, \tag{1}$$

where γ denotes the *risk attitude* of the agent; a higher γ implies more risk aversion. We shall assume that γ is known to the insurer, thereby eliminating adverse selection and solely focusing on the moral hazard aspect of the problem.

Using basic properties of the normal distribution, we have the following expected utility for the agent:

$$\overline{U}(e) = E(-\exp\{-\gamma \cdot (-L_e - ce)\}) = -\exp\{\gamma \cdot \mu(e) + \frac{1}{2}\gamma^2\lambda(e) + \gamma ce\}. \tag{2}$$

Using (2), the optimal effort for an agent outside the contract is given by $m := \arg\min_{e \ge 0} \{\mu(e) + \frac{1}{2}\gamma\lambda(e) + ce\}$. Let $U^o := \overline{U}(m)$ denote the maximum expected payoff of the agent without a contract.

Agent's Payoff with a Contract: If the agent accepts a contract, his utility is given by:

$$U^c(p, \alpha, \beta, e) = -\exp\{-\gamma \cdot (-p + \alpha \cdot S_e - L_e + \beta \cdot L_e - ce)\}.$$

Noting that S_e and L_e are conditionally independent, his expected utility is

$$\begin{aligned}\overline{U}^c(p, \alpha, \beta, e) &= E(-\exp\{-\gamma \cdot (-p + \alpha \cdot S_e - L_e + \beta \cdot L_e - ce)\}) \\ &= -\exp\left\{\gamma(p + (c-\alpha)e + \tfrac{1}{2}\alpha^2\gamma\sigma^2 + (1-\beta)\mu(e) + \tfrac{1}{2}\gamma(1-\beta)^2\lambda(e))\right\}\end{aligned}$$

The Insurer's Problem: The insurer designs the contract (p, α, β) to maximize her expected payoff. In doing so, the insurer also has to satisfy two constraints:

Individual Rationality (IR), and Incentive Compatibility (IC). The first stipulates that a rational agent will not enter a contract with payoff less than his outside option U^o, and the second that the effort desired by the insurer should maximize the agent's expected utility under that contract. Formally,

$$\max_{p,\alpha,0\leq\beta\leq1} \bar{V}(p,\alpha,\beta,e) = p - \alpha \cdot e - \beta \cdot \mu(e)$$

$$\text{s.t.} \quad \text{(IR)} \quad \bar{U}^c(p,\alpha,\beta,e) \geq U^o \tag{3}$$

$$\text{(IC)} \quad e \in \arg\max_{e'\geq0} \bar{U}^c(p,\alpha,\beta,e')$$

Note that the (IR) constraint can be re-written as follows,

$$p + (c-\alpha)\cdot e + \tfrac{1}{2}\alpha^2\cdot\gamma\sigma^2 + (1-\beta)\mu(e) + \tfrac{1}{2}\gamma(1-\beta)^2\lambda(e) \leq u^o .$$

where, $u^o := \frac{\ln(-U^o)}{\gamma} = \min_{e\geq0}\{\mu(e) + \tfrac{1}{2}\gamma\lambda(e) + c\cdot e\}$. Similarly, the (IC) constraint can be rearranged as follows,

$$e \in \arg\min_{e'\geq0} \ (c-\alpha)\cdot e' + (1-\beta)\mu(e') + \tfrac{1}{2}\gamma(1-\beta)^2\lambda(e').$$

3 The Role of Pre-screening in a Single Agent System

In this section, we first solve the optimization problem in (3). We then study the impact of several problem parameters, particularly the accuracy of pre-screening, on the optimal contract.

Lemma 1. *The (IR) constraint is binding in the optimal contract.*

By Lemma 1, an optimal contract satisfies the following equation:

$$p + (c-\alpha)\cdot e + \frac{1}{2}\alpha^2\cdot\gamma\sigma^2 + (1-\beta)\mu(e) + \frac{1}{2}\gamma(1-\beta)^2\lambda(e) = u^o .$$

We use the above expression to substitute for the base premium p in the objective function of (3), and re-writing the insurer's problem as follows,

$$\max_{\alpha,0\leq\beta\leq1,e\geq0} f(\beta,e,\alpha) = u^o - \mu(e) - \tfrac{1}{2}\gamma(1-\beta)^2\lambda(e) - c\cdot e - \tfrac{1}{2}\alpha^2\gamma\sigma^2$$
$$\text{s.t.,} \qquad e = \arg\min_{e'\geq0}(c-\alpha)\cdot e' + (1-\beta)\mu(e') + \tfrac{1}{2}\gamma(1-\beta)^2\lambda(e')$$
$$\tag{4}$$

We now turn to the issue of network security. We consider the effort level of the agent as the metric for evaluating the change in network security. We start with the following theorem on the state of network security, before and after the purchase of an insurance contract.

Theorem 1. *The effort exerted by the agent in the optimal contract is less than or equal to the level of effort outside the contract. In other words, insurance decreases network security as compared to the no-insurance scenario.*

Theorem 1 illustrates the inefficiency of cyber-insurance as a tool for improving the state of security. Existing work in [8,9] have also arrived at a similar conclusion when studying competitive/unregulated cyber-insurance markets. Nevertheless, as cyber-insurance is a profitable market, especially given risk-averse users, a market for cyber-insurance exists, and its growth is conceivable. We therefore ask whether the introduction of a pre-screening signal can lead to higher profits for the insurer, while also positively impacting the state of security, over the case of *no pre-screening*. We first analyze the impact of a pre-screening signal on the insurer's profit.

Theorem 2. *The insurer's payoff in the optimal contract increases as σ decreases. That is, the insurer's profit is increasing in the quality of the pre-screening signal.*

The above result is intuitively to be expected, as we predict that a strategic insurer can leverage the improved pre-screening information to her benefit, and attain better payoff. The more interesting observation is on the effect of pre-screening on the state of network security. The following theorem presents a sufficient condition under which the availability of a pre-screening signal improves network security, compared to the no pre-screening scenario. Note that we use $\sigma = \infty$ for evaluating the no pre-screening scenario. The equivalence follows from the fact that, as shown in [11], by setting $\sigma = \infty$, the insurer's optimal choice will be to set $\alpha = 0$, which effectively removes the effects of pre-screening.

Theorem 3. *Let e_1, e_2, e_∞ denote the optimal effort of the agent in the optimal contract when $\sigma = \sigma_1$, $\sigma = \sigma_2$ and $\sigma = \infty$, respectively. Let $k(e, \alpha) = \frac{\mu'(e) + \sqrt{\mu'(e)^2 - 2\gamma(c - \alpha)\lambda'(e)}}{-\gamma\lambda'(e)}$. If $k(e, \alpha_1)^2\lambda(e) - k(e, \alpha_2)^2\lambda(e)$ is non-decreasing in e for all $0 \leq \alpha_1 \leq \alpha_2 \leq c$, then $e_1 \geq e_2$ if $\sigma_1 \leq \sigma_2$. In other words, better pre-screening signals improve network security.*
In addition, if $k(e, 0)^2\lambda(e) - k(e, \alpha)^2\lambda(e)$ is non-decreasing in e for all $0 \leq \alpha \leq c$, then $e_1 \geq e_\infty$. In other words, availability of a pre-screening signal improves network security over the no pre-screening scenario.

In the above theorem, $k(e, \alpha)$ is in fact equivalent to $1 - \beta$. Consequently, $k(e, \alpha)^2\lambda(e)$ is the variance of the uncovered loss in a contract as a function (e, α). Therefore, Theorem 3 introduces a sufficient condition for improvement of network security based on the change in the variance of the uncovered loss.
Several instances of $\mu(e)$ and $\lambda(e)$ satisfy the condition of Theorem 3; for instance, $(\mu(e) = \frac{1}{e}, \lambda(e) = \frac{1}{e^2})$ or $(\mu(e) = \exp\{-e\}, \lambda(e) = \exp\{-2e\})$. Theorems 2 and 3 together imply that the introduction of a pre-screening signal benefits the insurer, as well the state of network security.

4   A Network of Two Risk Averse Agents

We next consider the one period contract problem between one risk-neutral insurer and two risk-averse agents. We assume the agents' utilities are again given by (1), and let γ_1, γ_2 denote the risk attitudes of the agents. We assume that the two agents are interdependent; the effort exerted by an agent affects not only himself, but further affects the loss that the other agent experiences. This assumption captures the fact that viruses, worms, etc., can spread from an infected agent to others. We model the interdependence between these two agents as follows,

$$L^{(i)}_{e_1,e_2} \sim \mathcal{N}(\mu(e_i + x \cdot e_{-i}), \lambda(e_i + x \cdot e_{-i}))$$

Here, $\{-i\} = \{1,2\} - \{i\}$, and $L^{(i)}_{e_1,e_2}$ is a random variable denoting the loss that agent i experiences, given both agents' efforts. The *interdependence factor* is denoted by x, and we let $0 \leq x < 1$.

The insurer can observe the result of pre-screening audit $S_{e_i} = e_i + W_i$ on each agent i, where W_i is a zero mean Gaussian noise with variance σ_i^2. We assume that W_1 and W_2 are independent, and that $S_{e_1}, S_{e_2}, L^{(1)}_{e_1,e_2}, L^{(2)}_{e_1,e_2}$ are conditionally independent given e_1, e_2.

We next separately analyze the following three cases, based on whether agents purchase cyber-insurance contracts.

(i) Neither agent enters a contract

(ii) One of the agents enters a contract, while the other one opts out

(iii) Both agents purchase contracts

Note that Case (ii) is the outside option for agents in Case (iii), and Case (i) is the outside option for agents in Case (ii). Therefore, in order to evaluate the participation constraints of agents when both purchase insurance contracts, we first need to find the optimal contract and agents' payoffs in Cases (i) and (ii).

4.1   Case (i): Neither Agent Enters a Contract

We start by considering the game G_{oo} between two agents, neither of which have purchased cyber-insurance contracts. The expected payoffs of these agents, with unit costs of effort $c_1, c_2 > 0$, are given by,

$$\bar{U}_i(e_1, e_2) = -\exp\{\gamma_i \mu(e_i + x \cdot e_{-i}) + \tfrac{1}{2}\gamma_i^2 \lambda(e_i + x \cdot e_{-i}) + \gamma_i \cdot c_i \cdot e_i\}$$

The best-response of each agent, when both opt out, can be found by solving the following optimization problem,

$$\begin{aligned}B_i^{oo}(e_{-i}) &= \arg\max_{e_i \geq 0} -\exp\{\gamma_i \mu(e_i + x \cdot e_{-i}) + \tfrac{1}{2}\gamma_i^2 \lambda(e_i + x \cdot e_{-i}) + \gamma_i \cdot c_i \cdot e_i\} \\ &= \arg\min_{e_i \geq 0} \mu(e_i + x \cdot e_{-i}) + \tfrac{1}{2}\gamma_i \lambda(e_i + x \cdot e_{-i}) + c_i \cdot e_i .\end{aligned}$$
$$(5)$$

The above optimization problem is a convex optimization problem and has a unique solution. In order to find $B_i^{oo}(e_{-i})$, we first define m_i as follows,

$$m_i := \arg\min_{e \geq 0}\{\mu(e) + \frac{1}{2}\gamma_i \lambda(e) + c_i \cdot e\} \tag{6}$$

Using (6), the solution to (5) is given by,

$$B_i^{oo}(e_{-i}) = \begin{cases} m_i - x \cdot e_{-i} & \text{if } m_i \geq x \cdot e_{-i} \\ 0 & \text{if } m_i \leq x \cdot e_{-i} \end{cases} \qquad (7)$$

The Nash equilibrium is given by the fixed point of the best-response mappings $B_1(e_2)$ and $B_2(e_1)$. Let $e_i^*(m_i, m_{-i})$ denote the effort of agent i at the *unique* Nash equilibrium. We have,

$$e_i^*(m_i, m_{-i}) = \begin{cases} \frac{m_i - x \cdot m_{-i}}{1 - x^2} & \text{if } m_i \geq x \cdot m_{-i} \text{ and } m_{-i} \geq x \cdot m_i \\ 0 & \text{if } m_i \leq x \cdot m_{-i} \\ m_i & \text{if } m_{-i} \leq x \cdot m_i \end{cases} \qquad (8)$$

Therefore, $\bar{U}_i^{*oo} = \bar{U}_i(e_1^*(m_1, m_2), e_2^*(m_2, m_1))$ is the utility of agent i in the equilibrium when agents do not choose to enter the contract. As we will see shortly, an insurer uses her knowledge of $\bar{U}_i^{*oo}$ to evaluate agents' outside options when proposing optimal contracts.

4.2 Case (ii): One of the Agents Enters a Contract

Assume that agent 1 enters a contract, while agent 2 opts out. We use G_{io} to denote the game between the insured agent 1 and uninsured agent 2. The expected payoffs of agents in this game are as follows,

$$
\begin{aligned}
U_1^{io}(e_1, e_2, p_1, \alpha_1, \beta_1) &= \\
&E(-\exp\{-\gamma_1 \cdot (-p_1 + \alpha_1 \cdot S_{e_1} - L_{e_1, e_2}^{(1)} + \beta_1 \cdot L_{e_1, e_2}^{(1)} - c_1 \cdot e_1)\}) \\
&= -\exp\{\gamma_1 \cdot (p_1 + (c_1 - \alpha_1) \cdot e_1 + \tfrac{1}{2}\alpha_1^2 \cdot \gamma_1 \sigma_1^2 \\
&\quad + (1 - \beta_1)\mu(e_1 + x \cdot e_2) + \tfrac{1}{2}\gamma_1(1 - \beta_1)^2 \lambda(e_1 + x \cdot e_2))\} \\
U_2^{io}(e_1, e_2) &= E(-\exp\{-\gamma_2(-L_{e_1, e_2}^{(2)} - c_2 \cdot e_2)\}) \\
&= -\exp\{\gamma_2\mu(e_2 + x \cdot e_1) + \tfrac{1}{2}\gamma_2^2 \lambda(e_2 + x \cdot e_1) + \gamma_2 \cdot c_2 \cdot e_2\}
\end{aligned}
$$

In order to find the Nash Equilibrium of G_{io}, we first calculate the best response of each agent. Let $B_i^{io}(e_{-i})$ denote the best response of agent i. We have,

$$
\begin{aligned}
B_1^{io}(e_2) &= \arg\max_{e_1 \geq 0} -\exp\{\gamma_1 \cdot (p_1 + (c_1 - \alpha_1) \cdot e_1 \\
&\quad + \tfrac{1}{2}\alpha_1^2 \cdot \gamma_1\sigma_1^2 + (1 - \beta_1)\mu(e_1 + x \cdot e_2) + \tfrac{1}{2}\gamma_1(1 - \beta_1)^2\lambda(e_1 + x \cdot e_2))\} \\
&= \arg\min_{e_1 \geq 0} (c_1 - \alpha_1) \cdot e_1 + (1 - \beta_1)\mu(e_1 + x \cdot e_2) + \tfrac{1}{2}\gamma_1(1 - \beta_1)^2\lambda(e_1 + x \cdot e_2)
\end{aligned} \qquad (9)
$$

As the above optimization problem is a convex problem, it has a unique solution. We next define $m_1(\alpha_1, \beta_1)$ as follows,

$$m_1(\alpha_1, \beta_1) = \arg\min_{e \geq 0}\{(c_1 - \alpha_1)e + (1 - \beta_1)\mu(e) + \frac{1}{2}\gamma_1(1 - \beta_1)^2\lambda(e)\}$$

Similar to (7), we use $m_1(\alpha_1, \beta_1)$ to find $B_1^{io}(e_2)$ as follows,

$$B_1^{io}(e_{-i}) = \begin{cases} m_1(\alpha_1, \beta_1) - x \cdot e_2 & \text{if } m_1(\alpha_1, \beta_1) \geq x \cdot e_2 \\ 0 & \text{if } m_1(\alpha_1, \beta_1) \leq x \cdot e_2 \end{cases} \qquad (10)$$

For the uninsured agent 2, it is easy to see that the best-response function is given by $B_2^{io}(e_1) = B_2^{oo}(e_1)$.

We can now find the Nash equilibrium as the fixed point of the best-response mappings. Agents' efforts at the equilibrium are $e_1^*(m_1(\alpha_1, \beta_1), m_2)$ and $e_2^*(m_2, m_1(\alpha_1, \beta_1))$ which are defined in (8). For notational convenience, we denote these efforts by e_1^*, e_2^*. Let $\bar{U}_i^{*io}$ denote the utility of agent i at effort levels e_1^*, e_2^*, in an equilibrium where only agent 1 purchases a contract, that is,

$$\bar{U}_1^{*io}(p_1, \alpha_1, \beta_1) = U_1^{io}(e_1^*, e_2^*, p_1, \alpha_1, \beta_1), \quad \bar{U}_2^{*io}(\alpha_1, \beta_1) = U_2^{io}(e_1^*, e_2^*)$$

Let $\bar{V}^{io}(p_1, \alpha_1, \beta_1, e_1, e_2)$ denote the insurer's utility, when she offers contract (p_1, α_1, β_1) to agent 1, and agents exert efforts e_1, e_2. The optimal contract offered by the insurer is the solution to the following optimization problem:

$$V^{*io} = \max_{p_1, \alpha_1, \beta, e_1^*, e_2^*} \bar{V}^{io}(p_1, \alpha_1, \beta_1, e_1^*, e_2^*) = p_1 - \alpha_1 e_1^* - \beta_1 \cdot \mu(e_1^* + x \cdot e_2^*)$$
s.t.,   (IR) $\bar{U}_1^{*io}(p_1, \alpha_1, \beta_1) \geq \bar{U}_1^{*oo}$,
(IC) e_1^*, e_2^* are effort of the agents in Nash equilibrium of game G_{io}

We first re-write the (IR) constraint for agent 1 as follows,

$$p_1 + (c_1 - \alpha_1) \cdot e_1^* + \frac{1}{2}\alpha_1^2 \gamma_1 \sigma_1^2 + (1 - \beta_1)\mu(e_1^* + x \cdot e_2^*) + \frac{1}{2}\gamma_1(1 - \beta_1)^2 \lambda(e_1^* + x \cdot e_2^*) \leq u_1^{oo},$$

where $u_1^{oo} = \frac{\ln(-\bar{U}_1^{*oo})}{\gamma_1}$.

Similar to Lemma 1, we can conclude that (IR) constraint is binding in the optimal contract. Therefore, we can re-write the insurer's problem by replacing for the base premium p, similar to the single agent problem in Sect. 2.

4.3   Case (iii): Both Agents Purchase Contracts

Assume the insurer offers each agent i a contract (p_i, α_i, β_i). The expected utility of agents when both purchase contracts is given by,

$$U_j^{(ii)}(e_1, e_2, p_j, \alpha_j, \beta_j) =$$
$$E(-\exp\{-\gamma_j \cdot (-p_j + \alpha_j \cdot S_{e_j} - L_{e_1, e_2}^{(j)} + \beta_j \cdot L_{e_1, e_2}^{(j)} - c_j \cdot e_j)\})$$
$$= -\exp\{\gamma_j \cdot (p_j + (c_j - \alpha_j) \cdot e_j + \frac{1}{2}\alpha_j^2 \cdot \gamma_j \sigma_j^2 + (1 - \beta_j)\mu(e_j + x \cdot e_{-j}) + \frac{1}{2}\gamma_j(1 - \beta_j)^2 \lambda(e_j + x \cdot e_{-j}))\}$$

Following steps similar to those in Sect. 4.2, the best-response function of player j, denoted B_j^{ii}, is given by,

$$B_j^{ii}(e_{-j}) = \begin{cases} m_j(\alpha_j, \beta_j) - x \cdot e_{-j} & \text{if } m_j(\alpha_j, \beta_j) \geq x \cdot e_{-j} \\ 0 & \text{if } m_j(\alpha_j, \beta_j) \leq x \cdot e_{-j} \end{cases}$$

where $m_j(\alpha_j, \beta_j)$ is the solution of the following equation,

$$m_j(\alpha_j, \beta_j) = \arg\min_{e \geq 0}(1 - \beta_j)\mu(e) + \frac{1}{2}\gamma_j(1 - \beta_j)^2 \lambda(e) + (c_j - \alpha_j) \cdot e. \quad (11)$$

Agents' efforts at the *unique* Nash equilibrium are $e_i^*(m_i(\alpha_i, \beta_i), m_{-i}(\alpha_{-i}, \beta_{-i}))$, with $e_i^*(.,.)$ defined in (8). For notational convenience, we simply denote these by e_i^*.

To write the insurer's problem, note that the outside option of agent 1 (resp. 2) from this game is the game G_{oi} (resp. G_{io}). The IR constraints can again be shown to be binding, simplifying the insurer's problem to,

$$V^{*ii} = \max\nolimits_{\alpha_1, 0 \le \beta_1 \le 1, \alpha_2, 0 \le \beta_2 \le 1, e_1^* \ge 0, e_2^* \ge 0} u_1^{oi} - \mu(e_1^* + x \cdot e_2^*)$$
$$-\tfrac{1}{2}\gamma_1(1-\beta_1)^2\lambda(e_1^* + x \cdot e_2^*) - c_1 \cdot e_1^* - \tfrac{1}{2}\alpha_1^2\gamma_1\sigma_1^2$$
$$+u_2^{io} - \mu(e_2^* + x \cdot e_1^*) - \tfrac{1}{2}\gamma_2(1-\beta_2)^2\lambda(e_2^* + x \cdot e_1^*) - c_2 \cdot e_2^* - \tfrac{1}{2}\alpha_2^2\gamma_2\sigma_2^2$$
$$s.t., e_1^*, e_2^* \text{ are the agents' effort in the equilibrium of game } G_{ii}$$

where $u_1^{oi} = \frac{\ln(-\bar{U}_1^{*oi})}{\gamma_1}$ and $u_2^{io} = \frac{\ln(-\bar{U}_2^{*io})}{\gamma_2}$, defined in Sect. 4.2.

5 The Role of Pre-screening in a Two Agent Network

We next discuss how different problem parameters, particularly the accuracy of pre-screening, affect the insurer's profit, as well as the system's state of security.

We first consider the utility of the insurer. As the insurer always has the option to not use the outcome of pre-screening by setting $\alpha = 0$ in the contract, the insurer's profit in the optimal contract with pre-screening is larger than her profit in the optimal contract without pre-screening; i.e., the availability of pre-screening is in the insurer's interest and improves insurer's profit.

We now return to the effect of pre-screening on the state of network security. We choose the total effort towards security, $e_1 + e_2$, as the metric for evaluating network security. The following two theorems characterize the impact of pre-screening on network security when the two agents are homogeneous ($\gamma_1 = \gamma_2 = \gamma, c_1 = c_2 = c, \sigma_1 = \sigma_2 = \sigma$). Theorem 4 shows that fully accurate pre-screening can improve network security over the no insurance scenario. Theorem 5 shows that under certain additional conditions, the improvement is still possible for sufficiently, yet not fully, accurate pre-screening.

Theorem 4. *Assume two homogeneous agents purchase (identical) contracts from an insurer, and let $m = \arg\min_{e \ge 0} \mu(e) + \frac{1}{2}\gamma\lambda(e) + ce$.*

(i) If $\mu'(m) < -\frac{c}{1+x}$ and both pre-screening signals are accurate, i.e., $\sigma_1 = \sigma_2 = 0$, then network security improves after the introduction of insurance.

(ii) If both of the pre-screening signals are uninformative, i.e., $\sigma_1 = \infty$ and $\sigma_2 = \infty$, network security worsens after the introduction of insurance.

Theorem 5. *Assume two homogeneous agents purchase (identical) contracts from an insurer. Let $m = \arg\min_{e \ge 0} \mu(e) + \frac{1}{2}\gamma\lambda(e) + ce$, $u_{max} = \mu(m) + \frac{1}{2}\gamma\lambda(m) + cm$, and $h(m', \beta) = c \cdot m' + (1-\beta)\mu(m') + \frac{1}{2}\gamma(1-\beta)^2\lambda(m')$. If $\mu'(m) < -\frac{c}{1+x}$, then there exists an upper bound $\sigma_{max}^2 := \min\{\frac{-\mu(m)-\frac{c \cdot m}{1+x}+\mu(0)}{0.5c^2\gamma}, \frac{-\mu'(m)-\frac{c}{1+x}}{M\gamma}\}$, where*

$$M := \max_{0 \le \beta \le 1, 0 \le m' \le \frac{(1+x)u_{max}}{c}} \{\frac{\partial h(m', \beta)}{\partial m'} \cdot \frac{\partial^2 h(m', \beta)}{\partial m'^2}\},$$

such that if $\sigma_1^2 = \sigma_2^2 \leq \sigma_{max}^2$, the existence of pre-screening improves network security as compared to the no insurance scenario.

6 Conclusion

We studied the problem of designing cyber-insurance contracts by a single profit-maximizing insurer, for both a single agent, as well as two interdependent agents. The introduction of insurance decreases network security in general, as agents reduce their effort after transferring part of their risks to an insurer. We propose the use of pre-screening signals on agents' efforts to prevent such reduction in effort after the introduction of insurance contracts, by offering premium discounts to agents with higher perceived efforts. We show that the availability of these pre-screening signals not only benefits the insurer by increasing her profit, but also improves network security, as compared to the no pre-screening scenario. Furthermore, when agents are interdependent and pre-screening is highly accurate, under a set of sufficient conditions, the incentivized improved efforts can increase network security not only over no pre-screening, but also compared to the no-insurance scenario. Therefore, introduction of pre-screening signals can be in the interest of the insurer, as well as the state of network security.

An important extension of this work is to consider arbitrary alternatives for including the pre-screening signals (as opposed to only linear discounts on premiums), and verify their role in improving network security. Considering multiple profit-maximizing insurers is another direction of future work.

Online Appendix. Numerical simulations and proofs are given in [11].

References

1. Bohme, R.: Cyber-insurance Revisited. Workshop on Economics Information Security (WEIS) (2005)
2. Bohme, R., Schwartz, G.: Modeling Cyber-Insurance: Towards a Unifying Framework. Workshop on Economics Information Security (WEIS) (2010)
3. Kesan, J., Majuca, R., Yurcik, W.: The economic case for cyberinsurance. In: Securing Privacy in the Internet Age. Stanford University Press (2005)
4. Kesan, J., Majuca, R., Yurcik, W.: Cyberinsurance as a Market-Based Solution to the Problem of Cybersecurity: a Case Study. Workshop on Economics Information Security (WEIS) (2005)
5. Hofmann, A.: Internalizing externalities of loss prevention through insurance monopoly. In: Proceeding Annual Meeting of American Risk and Insurance Association (2006)
6. Bolot, J., Lelarge, M.: Cyber-Insurance as an Incentive for Internet Security. Workshop on Economics Information Security (WEIS) (2008)
7. Pal, R., Golubchik, L., Psounis, K., Hui, P.: Will cyber-insurance improve network security? A market analysis. In: IEEE INFOCOM (2014)
8. Shetty, N., Schwartz, G., Felegyhazi, M., Walrand, J.: Competitive cyber-insurance and internet security. In: Economics of Information Security and Privacy (2010)

9. Shetty, N., Schwartz, G., Felegyhazi, M., Walrand, J.: Can competitive insurers improve network security?. In: International Conference on Trust and Trustworthy Computing (2010)

10. Osborne, M.J., Rubinstein, A.: A Course in Game Theory. MIT Press, Cambridge (1994)

11. www.dropbox.com/sh/euux09td56kqdnh/AAB6PGXFqa3BSbYyN4l1s5FZa?dl=0

A Game-Theoretic Model for Analysis and Design of Self-organization Mechanisms in IoT

Vahid Behzadan$^{(\boxtimes)}$ and Banafsheh Rekabdar

Department of Computer Science and Engineering,
University of Nevada, Reno, 1664 N Virginia Street, Reno, NV 89557, USA
{vbehzadan,brekabdar}@unr.edu

Abstract. We propose a framework based on Network Formation Game for self-organization in the Internet of Things (IoT). In this framework, heterogeneous and multi-interface nodes are modeled as self-interested agents who individually decide on establishment and severance of links to other agents. Through analysis of the static game, we formally confirm the emergence of realistic topologies from our model, and analytically establish the criteria that lead to stable multi-hop network structures.

Keywords: Internet of Things · Topology control · Self-organization · Game theory

1 Introduction and Motivation

Through the past decade, the number of internet-enabled devices has been growing at an unprecedented rate. The paradigm of Internet of Things (IoT) envisions an even more immersive and pervasive exploitation of internet connectivity by enabling more objects and devices to connect. Emerging applications of this move towards ubiquitous connectivity are wide and vast [1], ranging from domestic monitoring and smart home solutions to healthcare solutions [2], smart grids [3], and disaster monitoring [4]. It hence follows that instances of IoT will be comprised of a great number of various devices, each with unique requirements and capabilities, leading to heterogeneity both in terms of function and communications.

The inevitably high degree of heterogeneity and scalability of IoT, dim the odds of feasibility and scalability for centralized control approaches [5]. An alternative to centralized architectures for IoT are those that rely on autonomic management of connectivity and resources through self-configuration [6]. Such solutions model the network as a system comprised of individual agents, each of which aims to retain connectivity with the network while optimizing their objectives, such as energy consumption and throughput. Even though this multi-agent abstraction presents a promising approach towards scalability, the decentralized nature of self-configuring IoT gives rise to many critical challenges in mechanism design. Of the most critical of these challenges is the problem of topology control, which is further complicated by the heterogeneity of IoT devices. Owing

© ICST Institute for Computer Sciences, Social Informatics and Telecommunications Engineering 2017
L. Duan et al. (Eds.): GameNets 2017, LNICST 212, pp. 74–85, 2017.
DOI: 10.1007/978-3-319-67540-4_7

to the similarity of distributed IoT and Ad Hoc networks, the literature on self-organization and topological analysis of IoT are mainly focused on adopting techniques that are originally developed for generic distributed networks such as Wireless Sensor Networks (WSNs) [6]. Yet, unique features, such as the immense diversity in capabilities and requirements in all aspects of IoT present major distinguishing factors that necessitate the development of techniques specific to the challenges of this emerging technology (Fig. 1).

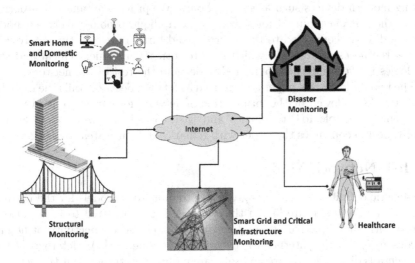

Fig. 1. Applications of IoT

The multi-agent model of IoT is comprised of opportunistic devices that aim to maximize their success in fulfilling their individual objectives, such as preservation of connectivity to the network, minimization of energy consumption and maintenance of a minimum Quality of Service (QoS). The inherent limitation of resources available to such opportunistic agents in any real-world deployment of IoT gives rise to a competitive environment, which motivates a game theoretic investigation of interactions in self-organizing IoT. The application of game theory to distributed topology control and self-organization has been an active area of research in recent years. Some of the notable literature in this area include the work of Eidenbenz et al. [7] on the analysis of equilibria in topology control games, Nahir et al.'s detailed investigation of applying game theory to various problems of topology control [8], and Saad et al.'s proposal of a game theoretic algorithm for cooperative relaying in [9], based on their earlier analysis of the formation of hierarchical topologies in multi-hop networks [10]. The models presented in these and many other topology control games, one critical limitation is the assumption on homogeneity of the network. Recently, Meirom et al. proposed a model of topology control games for heterogeneous AS-Level networks [11,12], which considers some degree of heterogeneity, but only accounts homogeneous link costs.

Based on the inevitable emphasis on the connectivity aspects of IoT networks, this paper builds on the aforementioned models to provide a framework for analysis and design of distributed topology control mechanisms in IoT. The proposed framework is based on modeling of self-organization as a Network Formation Game [13], in which the actions of players are establishment or severance of links with other nodes in the IoT. Contrary to previous models, we consider heterogeneity in both communications and link cost. The proposed model also accounts for nodes equipped with multiple communication interfaces, thus supporting modern devices such as smart phones. We provide an analytical derivation of the criteria required for formation of a clique topology between nodes that are directly connected to the internet, and further develop this analysis to present the necessary criteria which lead to formation of hierarchical and star topologies between internet-connected nodes and the rest of the network.

The remainder of this paper is organized as follows: Sect. 2 details the model of IoT networks, followed by the formulation of network formation game in Sect. 3. Emergence of stable IoT topologies and their criteria is discussed in Sect. 4. Finally, Sect. 5 concludes the paper with remarks on future areas of work.

2   IoT Network Model

The generic definition of IoT has given rise to numerous models for the network structure and architecture [5]. In this work, IoT is considered to be a network formed with the objective of enabling direct or relayed connectivity of heterogeneous nodes to the internet (or other backbone networks). Heterogeneity of nodes entails diverse hardware and software parameters throughout the network, such as the number and type of communication interfaces (e.g. WLAN, LTE, Ethernet, etc.), energy constraints, and bandwidth requirements.

Accordingly, we model the IoT as a network $G(P)$ of N nodes $P = \{P_i \ | \forall i \in \{1, 2, ..., N\}\}$, each with an arbitrary number of single channel radio interfaces. This definition may be seamlessly extended to cover multi-channel radios as well, via representing each as a group of single-channel radios. It is assumed that all interfaces of a node can be active simultaneously, but as detailed in Sect. 3, the effects of activating each additional interface on undesired aspects such as co- and cross-interference, channel congestion, and energy consumption may be suitably captured in the system cost function. The presented model also allows that some, or all of the interfaces in nodes may remain idle throughout the analyzed operation.

As the focus of this study is on topological properties, it is assumed that nodes are static relative to each other. Also, we consider the case that every node in the network is aware of its distance in terms of number of intermediate hops with every other node in the network. This can be justified by reliance on routing tables obtained from proactive network layer protocols such as OLSR [14]. The extent of a node's knowledge of the overall network topology is assumed to be limited to its directly connected neighbors.

Nodes are classified in two categories: Those with direct connectivity to the internet, such as WiFi Access Points and 3G/LTE Enabled Devices, and those which need to be connected to the internet via the nodes in the former group, such as Bluetooth/Zigbee sensors. Let the set of Internet Connected (IC) nodes $G_I \in P$ denote the set of nodes with direct connection to the internet, and the set of non-ICs $G_S \in P \backslash G_I$ is the set of nodes that do not have a direct connection to the internet. The emerging network is thus hierarchical with at least two tiers: a higher tier formed by IC nodes, and a lower tier comprised of non-IC nodes who aim to connect to the higher tier. Hence, an important objective of IoT network controllers, whether centralized or distributed, is to enable the connection to the internet to the non-IC node, via linking them to one or more IC nodes. In line with practical network protocols, a further limit is imposed to the maximum number of hops that may exist between each pair of nodes, denoted by h_{Max}. The following section provides the details of one such controller based on a game theoretical framework known as Network Formation Games.

3   Game Formulation

Formation of macro-scale topologies in distributed networks is the collective result of the individual decisions made by each nodes on which set of nodes to connect with, and which links to severe. With the assumption that every such node aims at gaining more utility from its decisions and consequent actions, this interaction of multiple decision makers can be formulated as a Network Formation Game [13]. Such games are comprised of competing agents who control the set of nodes they are connected to, with the common objective of forming coalitions of nodes that is most profitable for the deciding agent. It is evident that the game being considered in this work is of the non-cooperative type, since the decisions are made independently. Another assumption adopted in our proposal is that a link between two nodes is established if, and only if, both nodes consent to its establishment. This assumption emulates the real-world phenomenon that occurs in cost-optimizing distributed networks. A simple, yet realistic example is depicted in Fig. 2. This figure illustrates a network formation game in which the objective of all players is to minimize their cost while maintaining their reachability from any other player by at most one intermediate hop - a property that we shall label as one-hop-reachable. The cost incurred to each player of this game is the cost of establishing their immediate links (denoted by edge weights in Fig. 2), which is assumed to be the same for both of the linked nodes. If two nodes are not one-hop-reachable, their cost is set to be infinity. For instance, the cost incurred to node B is the cost of establishing the link BC plus the cost of establishing BD, i.e. $2 + 4 = 6$. As is shown in the figure, for node C to be one-hop-reachable to node A, the minimum cost is obtained by relaying through node D. Yet for node D, establishment of a link to node C does not bring any utility but losses, as node D has already established a cheaper path to C via node B, and is directly connected to node A. Hence, node D will not consent to spending its limited bandwidth and energy to relay a transmission that gains him no benefits. Consequently, nodes A and C settle on establishing an expensive direct link to avoid the infinite cost of unreachability.

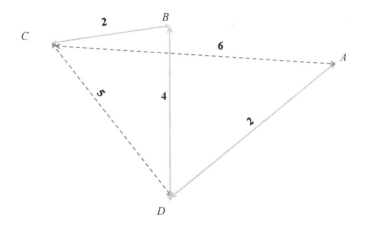

Fig. 2. Example of the mutual consent in Myerson games

Network formation games that are based on consensual establishment of links are known as bilateral linking or Myerson games [15], which is a widely adopted model in game theoretic distributed topology control, mainly due to its agreement with the opportunistic behavior of agents in decentralized networks. Our proposed framework builds atop of the previous work on bilateral link formation by extending the application of the Myerson model to considerations beyond that of minimizing energy consumption as the sole objective of the game, replacing the abstracted link establishment parameters with those of real wireless interface characteristics and propagation model, and filling the gap in self-organizing IoTs by providing a novel cross-layer framework for analytical design and evaluation of protocols and parameters involved in the distributed formation of IoT topologies.

Even though the real phenomenon of network formation in ad hoc communications networks is of a dynamic nature, this work concentrates on the analysis of a static bilateral linking game, with the aim of gaining insights on the characteristics of emerging stable topologies, along with the criteria that leads to their emergence. Similar to every other game, our proposed Myerson game is formed of players, set of strategies, and a payoff/cost structure, the details of each are presented in this section.

3.1 Players

Let $P = \{p_1, p_2, ...p_N\}$ denote a group of N agents. Each agent p_i is characterized by the following features:

- Ordered set of its radio interfaces R_i, where $|R_i|$ is the number of interfaces and $R_i(r \in \{1, 2, ..., |R_i|\}) \in \{0, 1\}$ is a binary value, indicating whether the interface is currently being used or not.
- Frequency of operation for each radio interface $f_{i,r}$

- Maximum bandwidth for each radio interface $b_{i,r}$
- Minimum required bandwidth b_i^{Max}
- Maximum transmit power for each radio interface $\tau_{i,r}$
- Receiver sensitivity for each radio interface $\S_{i,r}$
- Maximum antenna gain for each radio interface $x_{i,r}$
- 2-D Position $\gamma_i = (x_i, y_i)$
- Feature tuple for each interface $w_{i,r} = (f_{i,r}, b_{i,r}, b_i^{Max}, \tau_{i,r}, S_{i,r}, x_{i,r}, \gamma_i)$

Define the network topology $G = \{g_{ij} : i, k \in P, i \neq j\}$. If a bidirectional link is established between p_i and p_j, then $g_{ij} = (r_i, r_j)$, where $r_i \in R_i$ is the interface chosen by the node i to communicate with the corresponding interface in node j, i.e. $r_j \in R_j$. If there is no direct link between i and j, $g_{ij} = (-1, -1)$.

3.2 Strategies

Let C_i denote the cost function for every node p_i. Any node $p_i \in P$ may form a link $g_{ij} = (r_i, r_j)$ with any node p_j in the neighborhood $M(i)$, defined as the set of all nodes that fall within the maximum communications range of i, if:

1. Nodes must have at least one type of radio interface available and in common, i.e.:
2. $\Delta C(p_i, G + (r_i, r_j)) < 0$
3. $\Delta C(p_j, G + (r_i, r_j)) < 0$

Where $\Delta C(p_K, G + (r_k, r_l)) = C(p_k, G \cup \{(r_k, r_l)\}) - C(p_k, G)$ is the difference between the total cost to node p_k by establishing the link (r_k, r_l) and the total cost to p_k without the establishment of this link.

Agent p_i may remove a link with agent p_j in $M' \subset R_k^c\{R_i(k) \neq (-1, -1)\}$ if:

$$\Delta C(p_i, G - (r_i, r_j)) < 0$$

Where $\Delta C(p_k, G - (r_k, r_l)) = C(p_k, G \backslash \{(r_k, r_l)\}) - C(p_k, G)$ is the difference between the cost incurred by node p_k removing the link (r_k, r_l) and the cost incurred by maintaining this link.

3.3 Payoff Structure

For each node p_i, the payoff of forming a direct link is dependent on the set of objectives listed below:

1. Minimize the total cost of link establishment $\sum_{z=1}^{deg(p_i)} L_i'(z)$
2. Minimize the hop distance to all nodes in the network, with priority over minimizing distance to the nodes directly connected to the internet.
3. Minimize energy consumption by avoiding excessive relay transmissions

The corresponding cost function for each node is thus formulated as:

$$C(i, G) = C_i = \sum_{z=1}^{deg(p_i)} L_i'(z) + \Gamma \sum_{j \in G_I} h(i, j)$$
$$+ \sum_{k \in G_S} h(i, k) + B_i \qquad (1)$$

Where $L_i'(z)$ is the cost of establishing the z-th link of p_i, with $deg(p_i)$ denoting the number of links established by p_i. Let $L_i(z)$ be the link between nodes i and z. To model the link cost, the following factors are considered:

- $L_i(z)$ is directly proportional to the minimum transmission power required for z to receive the signal. The transmission power depends on the fading model and noise on the channel, which generally is inversely proportional to the Euclidean distance between nodes, their antenna gains, and the receiver sensitivity. Every interface has a maximum budgeted transmit power , beyond which $L_i(z) = \infty$
- $L - i(z)$ is directly proportional to the number of connections established on interface $R_i(r)$. The more this number is, the more congestion is expected and hence the throughput suffers.
- $L_i(z)$ is inversely proportional to bandwidth. Higher the bandwidth, higher the throughput will be.

Hence, a generic formulation for $L_i(z)$ is constructed as:

$$\sum_{z=1}^{deg(p_i)} L_i'(z) = \sum_{r=1}^{|R_i|} R_i(r) \sum_{z \in P s.t. g_{iz}=(R_i(r),o)} \alpha . \rho_i . \frac{\sigma_{ir}}{\beta_{ir}} \qquad (2)$$

Where α is a constant factoring the effect of each additional link on interface $R_i(r)$, ρ_i is the relative importance of preserving energy to achieving the desired throughput, σ_{ir} is the power transmitted by p_i on this link, and β_{ir} is the ratio of the available bandwidth to the required bandwidth, i.e.:

$$\beta_{ir} = \frac{b_{ir}}{b_i^{Min}} \qquad (3)$$

The factor $\Gamma \geq 1$ is the weighting factor for tuning the emphasize on minimizing the shortest hop-distance $h(i, j)$ to every IC node $j \in G_I$. B_i is the bridging coefficient of node p_i, estimating the local burden of bridging communities and thus modeling the relative amount of relay transmissions that p_i may have to handle for its neighbors. It is shown in [16,17] that the higher values of bridging coefficient represent a higher risk of congestion, as well as collisions. Bridging coefficient is calculated as:

$$B_i = \frac{\frac{1}{deg(i)}}{\sum_{j \in \{k : g_{ik} \neq (-1,-1)\}} \frac{1}{deg(j)}} \qquad (4)$$

4   Equilibrium Topologies in Static Game

This section investigates the criteria which enable the emergence of stable and efficient topologies from the proposed network formation mechanism. Having a game theoretic abstraction of the problem, we study the characteristics of stable networks by analyzing the equilibria of our model. One of the most intuitive types of equilibrium is the Nash equilibrium, defined as strategy profiles at which no player can increase its profit by unilaterally deviating from that profile, hinting at a stable outcome. Yet, Nash equilibrium is shown to be a weak notion for stability in network formation games [18]. Considering the bilateral nature of link formation in such games, stability of outcomes is characterized more accurately by considering bilateral deviations. To satisfy this requirement, we consider the notion of *pairwise stability* [18]. A strategy profile is said to be pairwise stable if no unilateral or bilateral deviations could increase the utility of the players. Formally, a topology G is pairwise stable if the following conditions are met:

1. $\forall i, ij \in G, C(i, G) \leq C(i, G - ij)$
2. $\forall i, j \notin G, if\ C(i, G + ij) < C(i, G)\ then\ C(j, G + ij) > C(j, G)$

In the following subsections, we utilize pairwise stability in the formal analysis of stable topologies that can emerge from the proposed model.

4.1   Formation of Cliques

A notable number of recent literature on bilateral link formation games are based on models that result in systematical limitation of pairwise stability to forest and tree topologies (e.g. [19,20]) This property greatly neuters the applicability of such models to IoT. As discussed in Sect. 2, nodes in IoT are categorized as either Internet-Connected (IC) or non-IC. It is intuitive to assume that each IC node is directly connected to every other IC nodes through the internet connection, thereby the set of all IC nodes inherently forms a clique. Therefore, if the cost of link establishment is bounded by a critical value, it is expected that the clique remains stable. In the following theorem, we prove that under certain criteria, this topology is indeed pairwise stable.

Theorem 1. *Let $L_i(k)$ be the maximum cost for any internet-connected node $p_i \in G_I$ to establish a link with node $p_k \in G_I$. If $L_i(k) < \Gamma - 1$, then the nodes in G_I form a clique.*

Proof. Assume a node p_i that is yet to establish connections to any node in G. For any node $p_k \in G_I$, the cost difference of establishing a link is given by:

$$C(p_i, G + g_{ik}) = C(p_i, G \cup \{r_i, r_k)\}) - C(p_k, G)$$
$$= L_i(k) + \Gamma(-1) + 0 + \Delta B_i \tag{5}$$

Where

$$\Delta B_i = \frac{\frac{1}{deg(i)+1}}{\sum_{j\in\{\forall\varsigma|g_{i\varsigma}\neq(-1,1)\}}\frac{1}{deg(j)}+1}$$
$$-\frac{\frac{1}{deg(i)}}{\sum_{j\in\{\forall\varsigma|g_{i\varsigma}\neq(-1,1)\}}\frac{1}{deg(j)}} \tag{6}$$

Considering the minimum and maximum values of $deg(i)$ and $\sum_{j\in\{\forall\varsigma|g_{i\varsigma}\neq(-1,1)\}}\frac{1}{deg(j)}\}$, it is trivial to show that:

$$0 < \Delta B_i < 1$$

Hence, the maximum valid value of the cost difference is given by:

$$\Delta C(p_i, G+g_{ik}) = L_i(k) - \Gamma + 1 \tag{7}$$

For this cost difference to be feasible for all nodes in G_I, the following condition must be satisfied:

$$\Delta C(p_i, G+g_{ik}) < 0$$
$$\Rightarrow L_i(k) - \Gamma + 1 < 0$$
$$\Rightarrow L_i(k) < \Gamma - 1 \tag{8}$$

If this condition holds true, establishment of a link between any pair of nodes in G_I decreases the cost for both nodes, hence leading to a clique topology. Inversely, severing any link in the resulting clique by any node $i \in G_I$ would impose a higher cost to i than gain. Therefore, this criteria leads to cliques that are pairwise-stable.

4.2   Formation of Stars and Hierarchies

Having established the criteria for the proposed model to result in a realistic stable topology for IC nodes, we study the topologies that emerge under this criteria for non-IC nodes. First, we derive the conditions that result in every non-IC node being linked to at most one of the IC nodes. Then, we derive the necessary conditions for formation of star clusters between non-IC nodes and IC nodes.

Theorem 2. *If $L_i(k) < \Gamma - 1$, the maximum number of links between any non-IC node $j \in G_S$ and the set of Internet-connected nodes G_I is 1.*

Proof. Assuming there already exists a link between $i \in G_I$ and $j \in G_S$, the maximum cost difference of establishing a second link from another node $i' \in G_I\backslash\{i\}$ to j is:

$$\Delta C(i', G+g_{i'j}) = L_{i'}(j) - \Gamma + 0 + 1 \tag{9}$$

For this link to be feasible for i', the following condition must be met:

$$\Delta C(i', G + g_{i'j}) < 0$$
$$\Rightarrow L_{i'}(j) < \Gamma - 1 \qquad (10)$$

Therefore, if the minimum cost of connection to a node $j \in G_S$ satisfies $L_{i'}(j) > \Gamma - 1$, every non-IC node is connected to at most one IC node.

In the following theorem, we derive the conditions under which every non-IC node is directly connected to an IC node, thus forming star-shaped clusters whose centers are IC nodes.

Theorem 3. *Let $L_i(k) < \Gamma - 1$ and $L_{i'}(j) > \Gamma - 1$, the maximum degree of any non-IC node $j \in G_S$ is 1 iff $\forall j' \in G_S \setminus j, L_j(j') > \frac{1}{2}$.*

Proof. Theorem 2 proves that under the aforementioned conditions, the maximum number of links between any non-IC node and all IC nodes is 1. Assume that j establishes is a second link to a node $j' \in G_S$. The cost difference is given by:

$$\Delta C(j, G + g_{jj'}) = L_j(j') + 0 - 1 + \frac{1}{2}$$
$$\qquad (11)$$

For this action to be infeasible, the cost difference must be positive. Therefore:

$$L_j(j') + 0 - 1 + \frac{1}{2} > 0$$
$$\Rightarrow L_j(j') > \frac{1}{2} \qquad (12)$$

As a corollary of Theorem 3, it is worth noting that if G is connected and the conditions of Theorems 1 and 2 are satisfied, but condition of Theorem 3 is not, then the resulting topology contains nodes that have one link to the IC set, but are connected to one or more non-IC nodes. Such nodes act as gateways and relays for other non-IC nodes connected to them, and the emerging topologies have more than the original 2 levels of hierarchy, namely IC and non-IC. Consequently, this model allows for resource planning by determination of nodes that are bound to become relays, and therefore require higher communications and processing capabilities.

5   Conclusions

In this paper, we proposed a model for self-organization in IoT based on bilateral link formation strategies. The model captures the heterogeneity of devices in IoT, as well as the emphasis on connectivity to the internet in the proposed cost function. The subsequent analysis of the static game established the criteria for emergence of cliques between the set of internet-connected nodes, as

well as multi-hop and star structures. Following the proposed model, further
analysis of the static game may provide insights into the efficiency of emerging
topologies, and establish the criteria for derivation of optimal network structures. Furthermore, this model provides a foundation for design and evaluation
of dynamic games and algorithms for distributed self-organization in heterogeneous networks such as IoT.

References

1. Bandyopadhyay, D., Sen, J.: Internet of Things: applications and challenges in
   technology and standardization. Wirel. Pers. Commun. **58**(1), 49–69 (2011)
2. Rohokale, V.M., Prasad, N.R., Prasad, R.: A cooperative Internet of Things (IoT)
   for rural healthcare monitoring and control. In: 2011 2nd International Conference on Wireless Communication, Vehicular Technology, Information Theory and
   Aerospace & Electronic Systems Technology (Wireless VITAE), pp. 1–6. IEEE
   (2011)
3. Yun, M., Yuxin, B.: Research on the architecture and key technology of Internet of
   Things (IoT) applied on smart grid. In: 2010 International Conference on Advances
   in Energy Engineering (ICAEE), pp. 69–72. IEEE (2010)
4. Fang, S., Da Xu, L., Zhu, Y., Ahati, J., Pei, H., Yan, J., Liu, Z.: An integrated
   system for regional environmental monitoring and management based on Internet
   of Things. IEEE Trans. Ind. Inform. **10**(2), 1596–1605 (2014)
5. Ma, H., Liu, L., Zhou, A., Zhao, D.: On networking of Internet of Things: explorations and challenges (2015)
6. Athreya, A.P., Tague, P.: Network self-organization in the Internet of Things. In:
   2013 IEEE International Conference on Sensing, Communications and Networking
   (SECON), pp. 25–33. IEEE (2013)
7. Eidenbenz, S., Kumar, V., Zust, S.: Equilibria in topology control games for ad
   hoc networks. Mob. Netw. Appl. **11**(2), 143–159 (2006)
8. Nahir, A., Orda, A., Freund, A.: Topology design and control: a game-theoretic
   perspective. In: INFOCOM 2009, pp. 1620–1628. IEEE (2009)
9. Saad, W., Han, Z., Basar, T., Debbah, M., Hjorungnes, A.: Network formation
   games among relay stations in next generation wireless networks. IEEE Trans.
   Commun. **59**(9), 2528–2542 (2011)
10. Saad, W., Zhu, Q., Basar, T., Han, Z., Hjorungnes, A.: Hierarchical network for-
    mation games in the uplink of multi-hop wireless networks. In: Global Telecom-
    munications Conference (GLOBECOM 2009), pp. 1–6. IEEE (2009)
11. Meirom, E.A., Mannor, S., Orda, A.: Strategic formation of heterogeneous net-
    works. arXiv preprint arXiv:1604.08179 (2016)
12. Meirom, E.A., Mannor, S., Orda, A.: Formation games of reliable networks. In:
    2015 IEEE Conference on Computer Communications (INFOCOM), pp. 1760–
    1768. IEEE (2015)
13. Jackson, M.O.: A survey of network formation models: stability and efficiency. In:
    Group Formation in Economics: Networks, Clubs, and Coalitions, pp. 11–49 (2005)
14. Clausen, T., Jacquet, P.: Optimized link state routing protocol (OLSR). Technical
    report (2003)
15. Dutta, B., Jackson, M.O.: Networks and Groups: Models of Strategic Formation.
    Springer, Heidelberg (2013). doi:10.1007/978-3-540-24790-6

16. Hwang, W., Kim, T., Ramanathan, M., Zhang, A.: Bridging centrality: graph mining from element level to group level. In: Proceedings of the 14th ACM SIGKDD International Conference on Knowledge Discovery and Data Mining, pp. 336–344. ACM (2008)
17. Komali, R.S., MacKenzie, A.B.: Distributed topology control in ad-hoc networks: a game theoretic perspective. In: Proceedings of IEEE CCNC, pp. 563–568 (2006)
18. Bloch, F., Jackson, M.O.: Definitions of equilibrium in network formation games. Int. J. Game Theor. **34**(3), 305–318 (2006)
19. Arcaute, E., Dallal, E., Johari, R., Mannor, S.: Dynamics and stability in network formation games with bilateral contracts. In: 2007 46th IEEE Conference on Decision and Control, pp. 3435–3442. IEEE (2007)
20. Arcaute, E., Johari, R., Mannor, S.: Network formation: bilateral contracting and myopic dynamics. IEEE Trans. Autom. Control **54**(8), 1765–1778 (2009)

A Dynamic Incentive Mechanism for Security in Networks of Interdependent Agents

Farzaneh Farhadi[1,2]([✉]), Hamidreza Tavafoghi[1], Demosthenis Teneketzis[1], and Jamal Golestani[2]

[1] University of Michigan, Ann Arbor, USA
{ffarhadi,tava,teneket}@umich.edu
[2] Sharif University of Technology, Tehran, Iran
golestani@ieee.org

Abstract. We study a dynamic mechanism design problem for a network of interdependent strategic agents with coupled dynamics. In contrast to the existing results for static settings, we present a dynamic mechanism that is incentive compatible, individually rational, budget balanced, and social welfare maximizing. We utilize the correlation among agents' states over time, and determine a set of *inference signals* for all agents that enable us to design a set of incentive payments that internalize the effect of each agent on the overall network dynamic status, and thus, align each agent's objective with the social objective.

Keywords: Security games · Dynamic mechanism design · Epidemics over networks

1 Introduction

Recently there has been a growing body of literature studying the dynamic behavior of networked strategic agents, where each agent's state and utility is affected by his interactions with his neighbors in the network. This literature is motivated by various applications that include opinion dynamics in social networks, epidemics spreading in networks, dynamic adoption of new products and technologies over networks, and network security. In this paper, we study a model of dynamic networked agents motivated by a network security application.

We consider a dynamic network with strategic agents who privately observe their own security state and are only interested in maximizing their own utility. We formulate a mechanism design problem for a network manager whose objective is to dynamically allocate his limited security resources in the network so as to maximize the overall security of the whole network over time.

We assume that an agent's utility depends on his own private security state as well as the externality he receives from his neighbors in the network. Moreover, an agent's security state dynamically evolves over time; its evolution depends

This work was supported in part by the NSF grants CNS-1238962 and CCF-1111061.

L. Duan et al. (Eds.): GameNets 2017, LNICST 212, pp. 86–96, 2017.
DOI: 10.1007/978-3-319-67540-4_8

on the security resources the agent receives from the network manager, as well as direct external attacks launched from outside of the network and indirect internal attacks launched from his unsafe local neighbors in the network. Therefore, the network manager needs to design a dynamic incentive mechanism for agents with *correlated types* and *interdependent valuations* so as to align their selfish objectives with his own objective, which is the maximization of the overall security of the whole network.

We propose a dynamic incentive mechanism that is individually rational and budget balanced [3], and enables the network manager to achieve the socially efficient outcome. Our result is in contrast with the existing impossibility results for incentive mechanisms that are socially efficient, individually rational, incentive compatible, and budget balanced in the static settings [22]. We exploit the dynamic correlation among the agents' security states and determine a set of *inference signal*s for all agents over time. Utilizing the proposed set of inference signals, we characterize a dynamic incentive mechanism that ensures the agents' incentive compatibility and individual rationality, achieves a socially efficient outcome, and is ex-ante budget balanced.

There is a growing body of literature on network security games (see [18] and references therein). One set of papers assume that network agents are cooperative, and study the interactions between the network as a whole and an outside attacker as a two-player attacker-defender game [4,13,17]. Another set of papers assume an exogenously-fixed attack behavior from outside the network, and study the interactions between strategic agents within the network as a network game problem (see [9,15] and references therein). For instance, the work of [10] studies a network security game with strategic agents, and shows that the equilibrium outcome of the game can be very poor compared to the social optimum, and this gap tends to increase with the increase in network size and the agents' interdependence. In our work, we study the dynamic interactions among agents within the network. However, we take the mechanism design approach rather than analyzing the resulting security game for a given environment.

The existing literature on mechanism design for network security considers mainly static incentive design problems. For instance, the work of [16] investigates the role of cyber-insurance as an incentive instrument for agents to increase their security investment in self-protection. The work of [22] studies the mechanism design problem for general networks with strategic agents in static settings, and shows that there exists no incentive mechanism that can implement the socially efficient outcome, while ensuring individual rationality, incentive compatibility and (weak) budget balance. Our paper contributes to this set of literature by showing that this impossibility result does not hold for dynamic settings. The fact that the agents' incentive problem improves in dynamic settings has been previously shown by works that look at security games in repeated settings (see [10,21]). Our work is different from those that consider repeated game settings. First, we take a mechanism design approach rather than analyzing a repeated game setting. Second, in repeated game settings there is no system dynamics, and the existing results are based on the reputation that agents mutually form over time. Our work provides another insight for such improvements in dynamic settings by capitalizing on the coupling among the agents' security dynamics over time.

The model we consider in this paper is also related the literature on Susceptible-Infected-Susceptible (SIS) epidemic models over networks with strategic agents (see [23] and references therein). For instance, the works of [8,24] study different variations of SIS epidemic models over networks with strategic agents from a game-theoretical approach. The authors in [24], investigate a game setting where agents make one-time investment decisions in their security which then affect the epidemic process. The work of [8] studies a marketing problem on networks using a SIS epidemic model, and investigates a game problem between two firms which compete for market shares over the network.

The mechanism design problem we consider in this paper can also be viewed as a dynamic resource allocation mechanism with strategic agents. The work of [12] studies the resource allocation problem in networks with non-strategic agents. The authors in [6,11] consider the resource allocation problem in static networks with strategic agents, take an *implementation theory* approach, and propose resource allocation mechanisms that are social welfare maximizing, individually rational and budget balanced. In this paper, we consider a class of dynamic resource allocation problems with strategic agents, and we present a dynamic resource allocation mechanism that is social welfare maximizing, incentive compatible, individually rational and ex-ante budget balanced.

The rest of the paper is organized as follows. We present our model in Sect. 2. We formulate the dynamic incentive design problem and characterize its solution in Sect. 3. We show that the dynamic incentive mechanism proposed in this paper can implement the solution of the corresponding dynamic centralized optimal resource allocation problem. In Sect. 4, we formulate such a centralized resource allocation problem as a centralized stochastic control problem and provide its solutions for a set of specific network topologies. The proofs of all the results that appear in this paper can be found in [7].

2   Model

There are n strategic agents each one residing in a distinct node of an interconnected network interacting over time $t \in \mathcal{T} := \{0, 1, 2, \ldots\}$. At each time $t \in \mathcal{T}$, the security state of agent i is given by $\theta_t^i \in \Theta := \{0, 1\}$; the realization of θ_t^i is agent i's private information. Agent i's state is safe if $\theta_t^i = 1$ and is unsafe if $\theta_t^i = 0$. We refer to θ_t^i as agent i's *type* at time t. There is a network manager who takes security measures dynamically over time so as to defend the network against external attacks and/or propagation of internal attacks. The security state θ_t^i of agent i dynamically evolves over time; θ_t^i's evolution depends on the security state of his neighbors in the network, the network manager's actions, and the probability of external attacks.

System Dynamics. We represent the agents' network by a directed graph $G = (N, L)$ where $N = \{1, ..., n\}$ and $L \in \mathbb{R}_+^{n \times n}$ denote the set of agents and the set of directed links between them, respectively. The state θ_t^i of agent i

is affected by agent j if $l_{ji} > 0$. We define the set of agent i's neighbors as $N^i := \{j : l_{ji} > 0\}$. During each time $t \in \mathcal{T}$, if agent i is in the safe state, i.e. $\theta_t^i = 1$, it may be attacked directly from outside with probability d_i, or indirectly from any of his unsafe neighbors $j \in N^i$ with probability l_{ji}. The topology of the network G and the probability of outside attacks d_i remains the same over time.

The goal of the network manager is to maximize the overall security of the network over time, i.e. maximize the social welfare. At each time t, the manager can choose one agent $a_t \in N$ and apply a security measure to him. As a result of applying the security measure to agent i, i.e. $a_t = i$, if agent i is in the unsafe state he will switch to the safe state with probability h. The security measure also protects the chosen agent against direct attacks from outside during time t with the same probability h, but it does not affect the indirect spread of attacks within the network.

Let $\boldsymbol{\theta}_t = (\theta_t^1, \ldots, \theta_t^n) \in \Theta^n$ denote the security state of the network at time t. As a result of the network manger's action a_t, new direct attacks from outside, and the spread of indirect attacks within the network during time t, the network state $\boldsymbol{\theta}_{t+1}$ has the following Markovian dynamics:

$$\mathbb{P}\{\boldsymbol{\theta}_{t+1} = \boldsymbol{b} | \boldsymbol{\theta}_t, a_t\} = \prod_{i=1}^n \mathbb{P}\{\theta_{t+1}^i = b_i | \boldsymbol{\theta}_t, a_t\}, \ \forall \boldsymbol{b} \in \Theta^n, \tag{1}$$

where,

$$\mathbb{P}\{\theta_{t+1}^i = 1 | \boldsymbol{\theta}_t, a_t\} = \begin{cases} 0, & \theta_t^i = 0, i \neq a_t \\ h(1 - d_i(1 - h)) \prod_{j \in N^i : \theta_t^j = 0} (1 - l_{ji}), & \theta_t^i = 0, i = a_t \\ (1 - d_i) \prod_{j \in N^i : \theta_t^j = 0} (1 - l_{ji}), & \theta_t^i = 1, i \neq a_t \\ (1 - d_i(1 - h)) \prod_{j \in N^i : \theta_t^j = 0} (1 - l_{ji}), & \theta_t^i = 1, i = a_t \end{cases}, \tag{2}$$

and $\mathbb{P}\{\theta_{t+1}^i = 0 | \boldsymbol{\theta}_t, a_t\} = 1 - \mathbb{P}\{\theta_{t+1}^i = 1 | \boldsymbol{\theta}_t, a_t\}$. We note that by (1) and (2) we assume that the outside attacks and attack spreads within network are independent across different agents, and thus, conditioned on previous state $\boldsymbol{\theta}_t$ and the network manager's action a_t, the agents' security states evolve independently as in (1). Equation (2) describes this evolution: (i) if agent i is in the unsafe state and is not receiving any security measure from the network manager at t, he remains in the unsafe state; (ii) if agent i is in the unsafe state and receives the security measure from the network manager, he will restore his security if the security measure is successful (prob. h), he is not the subject of new direct attacks (prob. $(1 - d_i(1 - h))$), and he is not attacked by his unsafe neighbors (prob. $\prod_{j \in N^i : \theta_t^j = 0} (1 - l_{ji})$); (iii) similarly, if agent i is in the safe state and is not receiving a security measure, he will remain in the safe state if he is not attacked from outside (prob. $1 - d_i$) and he is not attacked by his unsafe neighbors (prob. $\prod_{j \in N^i : \theta_t^j = 0} (1 - l_{ji})$); (iv) if agent i is in the safe state and is receiving a security measure from the network manager, he will remain in the safe state if he is not attacked from outside (prob. $1 - d_i(1 - h)$) and he is not attacked by his neighbors that are in an unsafe state (prob. $\prod_{j \in N^i : \theta_t^j = 0} (1 - l_{ji})$).

Agents' Utilities. Each agent $i \in N$ has a valuation for his security state θ_t^i as well as the security state of his neighbors θ_t^j, $j \in N^i$, and the security measures he receives from the network manager; this valuation is given by,

$$v^i(\boldsymbol{\theta}_t, a_t) = \theta_t^i + \frac{\alpha}{|N^i|} \mathbb{1}_{\{\theta_t^i = 1 \text{ or } a_t = i\}} \sum_{j \in N^i} \theta_t^j, \tag{3}$$

where $0 < \alpha < 1$ captures the value of a safe neighborhood to an agent i. As a result of (3), agent i has a positive valuation for safe neighbors only if he is in the safe state or he is receiving a security measure at t, i.e. $\{\theta_t^i = 1 \text{ or } a_t = i\}$. Let p_t^i denote the monetary payment made by agent i to the network manager at t $(p_t^i \in \mathbb{R})$. Then the total utility of agent i at t is given by,

$$u_t^i(\boldsymbol{\theta}_t, a_t, p_t^i) = v^i(\boldsymbol{\theta}_t, a_t) - p_t^i, \tag{4}$$

Let $\delta \in (0, 1)$ denote the common discount factor. Then the total discounted utility of agent $i \in N$, is

$$U^i = (1 - \delta) \sum_{t=0}^{\infty} \delta^t u_t^i(\boldsymbol{\theta}_t, a_t, p_t^i) = (1 - \delta) \sum_{t=0}^{\infty} \delta^t (v^i(\boldsymbol{\theta}_t, a_t) - p_t^i). \tag{5}$$

The network manager's objective is to maximize the social welfare W given by,

$$W = \mathbb{E}\{(1 - \delta) \sum_{t=0}^{\infty} \delta^t \sum_{i=1}^{n} v^i(\boldsymbol{\theta}_t, a_t)\}. \tag{6}$$

The network manager's problem would be a standard control problem (Markov decision problem) if the manager knew $\boldsymbol{\theta}_t$ for all t. However, $\boldsymbol{\theta}_t$ is not known to the manager; θ_t^i, $i \in N$, is agent i's private information. Thus, in order to take a security measure at any time t, the manager has to elicit information about each agent's security status. Since all agents are selfish (strategic) and want to maximize their own utility given by (5), they do not voluntarily reveal their information to the manager. Therefore, the manager needs to design an incentive mechanism so as to align the agents' objectives with his own objective. In this paper, we investigate such an incentive design problem, and formulate it as a mechanism design problem in Sect. 3.

3   Dynamic Incentive Design Problem

We invoke the revelation principle for dynamic games [20], and, without loss of generality, restrict attention to *direct revelation mechanisms* that are *incentive compatible*. In a direct revelation mechanism, at every $t \in T$, the network manager asks agents to report their current security state. Let r_t^i denote agent i's report for time t, which is not necessarily the same as θ_t^i. Let $h_t := \{r_s^i, i \in N, s \leq t\}$ denote the history of reports and $\mathcal{H}_t$ denote the set of all possible histories at t. A direct mechanism is captured by a set of functions $(\pi(.), p(.)) = \{\pi_t(\cdot), p_t^i(\cdot), i \in N, t \in T\}$ that the network manager designs and commits to them, where $\pi_t : \mathcal{H}_t \to N$ determines which agent receives the security measure at t, and $p_t^i : \mathcal{H}_t \to \mathbb{R}$, $i \in N$, determines the monetary payment (or the negative of the monetary incentive) that agent i makes (receives) at time t based on the history up to t. A direct mechanism is incentive compatible (IC) if at every $t \in \mathcal{T}$ every agent is willing to report truthfully his security state given that the other agents report truthfully. That is, for every agent $i \in N$ and for all reporting strategies $\{\sigma_\tau^i : \Theta \times \mathcal{H}_\tau \to \Delta(\Theta), \tau \geq t\}$, truth telling results in higher expected utility at every $t \in \mathcal{T}$ and $h_t \in \mathcal{H}_t$, i.e.

$$
\begin{aligned}
&\mathbb{E}\{(1-\delta) \sum_{\tau=t}^{\infty} \delta^{\tau-t} \left[v^i(\boldsymbol{\theta}_\tau, \pi_\tau(\boldsymbol{\theta}_\tau^{-i}, \theta_\tau^i)) - p_\tau^i(\boldsymbol{\theta}_\tau^{-i}, \theta_\tau^i) \right]\} \geq \\
&\mathbb{E}\{(1-\delta) \sum_{\tau=t}^{\infty} \delta^{\tau-t} \left[v^i(\boldsymbol{\theta}_\tau, \pi_\tau(\boldsymbol{\theta}_\tau^{-i}, \sigma_\tau(\theta_\tau^i, h_\tau^i))) - p_\tau^i(\boldsymbol{\theta}_\tau^{-i}, \sigma_\tau(\theta_\tau^i, h_\tau^i)) \right]\},
\end{aligned}
\tag{7}
$$

where $\Delta(\Theta)$ denotes the set of all probability distributions on Θ.

The network manager also needs to ensure that agents voluntarily participate in the direct mechanism $(\pi(.), p(.))$. Let $U_0^i \geq 0$ denote agent i's expected utility by opting out of the mechanism. Then, agents' voluntary participation is ensured by the following individual rationality (IR) constraints as follows,

$$
\mathbb{E}\{(1-\delta) \sum_{\tau=0}^{\infty} \delta^\tau \left[v_\tau^i(\boldsymbol{\theta}_\tau, \pi_\tau(\boldsymbol{\theta}_\tau^{-i}, \theta_\tau^i)) - p_\tau^i(\boldsymbol{\theta}_\tau^{-i}, \theta_\tau^i) \right]\} \geq U_0^i, \forall i \in N.
\tag{8}
$$

Therefore, we can formulate the dynamic incentive design problem for the network manager as follows:

$$
\max_{\pi(\cdot), p(\cdot)} \mathbb{E}\{(1-\delta) \sum_{t=0}^{\infty} \delta^t \sum_{i=1}^{n} v^i(\boldsymbol{\theta}_t, a_t)\}
\tag{9}
$$

subject to IC constraints (7) and IR constraints (8)

The incentive design problem formulated above is a dynamic mechanism design problem with correlated types and interdependent valuations. It is a dynamic mechanism design (in the strategic sense) since agents' incentive constraints at any time t depend on their strategic decisions at other times. Moreover, since the evolution of security states, given by (2), are coupled among agents, the agents' types are correlated with each other and over time. Furthermore, each agent's utility, given by (3), depends on his neighbor's security states in addition to his own security state, thus, agents have interdependent valuations. As a result of the correlation among agents' types and agents' interdependent valuations, the dynamic generalizations of the Vickrey–Clarke–Groves (VCG) mechanism [2] and that of d'Aspremont and Gerard-Varet (AGV) mechanism [1] cannot be used to solve the network manager's problem (9).

In this paper, we present an alternative approach to the dynamic incentive design problem by the network manager. We utilize the correlation among agents' security states over time to form a set of *cross inference signals* that enable us to internalize the effect of each agent's security state on the overall network security through incentive payments. The idea of utilizing the correlation among agents' types to extract their private information was first exploited by Cremer and McLean in a static setting [5]. They formed a cross inference signal for each agent by utilizing the correlation among the realization of agents' types, determined appropriate incentive payments that depend on the cross inference signals, and extracted the agents' private information. Liu [19] considered a dynamic setting with coupled dynamics, and utilized the inter-temporal correlation among agents' types to form cross inference signals for each agent that lead to truthful reporting at each time instant.

We provide a similar approach as the one in [19]. We utilize the inter-temporal correlation between agent i's security state θ_t^i at t and other agents' security state θ_{t+1}^j, $j \neq i$, at $t+1$ and form a cross inference signal that determines agent i's payment over time. We show that such cross inference signals enable the network manager to align the agents' self-interests with the overall social interest, and maximize the social welfare W.

3.1   Specification of the Mechanism

In this section we present a 'Dynamic Cross Inference' (DCI) mechanism that maximizes the social welfare subject to the IC and IR constraints (9). The description of our mechanism is divided into two parts: the allocation policy $\{\pi_t(\cdot), t \in \mathcal{T}\}$, and the monetary transfers $\{p_t^i(\cdot), i \in N, t \in \mathcal{T}\}$.

Allocation Policy. The specification of the allocation policy is based on the premise that the mechanism is incentive compatible. In an incentive compatible mechanism the agents report their security states truthfully. Therefore, the network manager is faced with a stochastic control problem with complete information. We design the allocation policy of our mechanism to be an optimal solution to this problem which we denote by π^*, i.e., $\pi_t = \pi^*(r_t)$, $\forall t \in \mathcal{T}$. In Sect. 4, we discuss how the network manager can find such an optimal policy.

Monetary Transfers. To obtain an incentive compatible mechanism, we design monetary transfers so that they exactly align the incentives of each agent with the social welfare. Since agents' valuations are interdependent, we cannot use the idea of Groves' mechanism by simply paying each agent i the total valuations of other agents, because the valuations of agents except i depend directly on the report of agent i, and this creates incentive for misreporting. To fix this, we utilize the correlation between agent i's security state θ_t^i at t and other agents' security states θ_{t+1}^j, $j \neq i$, at $t+1$ and form a cross inference signal about the security state of agent i which is independent of his own reports. We use this cross inference signal to align the objective of agent i with the social welfare.

Specifically, let r_t^{-i} denote the report profile of all agents except agent i at time t. We define the cross inference signal for agent i at time t as follows:

$$m_t^i = \begin{cases} 0, & \text{if } r_{t+1}^j = 0, \forall j \in O^i, \\ 1, & \text{otherwise,} \end{cases} \tag{10}$$

where $O^i := \{j \in N : i \in N^j\}$ is the set of *output neighbors* of agent i. If at time $t+1$, all output neighbors of agent i report to be unsafe, the manager interprets this as a signal that agent i was unsafe at time t. Otherwise, he assesses agent i as a safe agent.

By using the cross inference signal m_t^i, we construct payments p_{t+1}^i such that, in expectation, at time $t+1$ agent i receives the sum of time-t flow valuations of all other agents. So agent i's continuation payoff at time t is equal to the social surplus from time t onward. With this in mind, we define the tax $p_{t+1}^i(m_t^i, \boldsymbol{r}_t^{-i}, a_t)$ to be paid by each agent i at time $t+1$, as the solution to the following system of linear equations:

$$\mathcal{P}(m_t^i = 0 | \theta_t^i, \boldsymbol{r}_t^{-i}, a_t) p_{t+1}^i(0, \boldsymbol{r}_t^{-i}, a_t) + \mathcal{P}(m_t^i = 1 | \theta_t^i, \boldsymbol{r}_t^{-i}, a_t) p_{t+1}^i(1, \boldsymbol{r}_t^{-i}, a_t) = \\ -\frac{1}{\delta} \sum_{j \neq i} v^j(\boldsymbol{\theta}_t, a_t), \forall \theta_t^i \in \Theta, \quad (11)$$

where $\mathcal{P}(m_t^i | \theta_t^i, \boldsymbol{r}_t^{-i}, a_t)$ is the probability of m_t^i given θ_t^i, $\boldsymbol{r}_t^{-i}$ and a_t, assuming truthful reports of agents except i, i.e. $\boldsymbol{r}_\tau^{-i} = \boldsymbol{\theta}_\tau^{-i}$, $\tau = t, t+1$.

Lemma 1. For any a_t and $\boldsymbol{r}_t^{-i}$, the system of equations (11) has a solution.

Therefore, payments p_{t+1}^i are always well-defined. Using these payments the network manager is able to align the objective of each agent with the social welfare since,

$$v^i(\boldsymbol{\theta}_t, a_t) - \delta \, \mathbb{E}\{p_{t+1}^i(m_t^i, \boldsymbol{\theta}_t^{-i}, a_t)\} = \sum_{j \in N} v^j(\boldsymbol{\theta}_t, a_t). \quad (12)$$

This feature is the key to proving the main result of this paper stated below.

Theorem 1. The DCI mechanism maximizes the social welfare and satisfies the IC and IR constraints, therefore, it is an optimal solution to the dynamic incentive design problem (9) for the network manager.

3.2   Budget Balance

The DCI mechanism proposed in Sect. 3.1 efficiently solves the problem network manager faces (9), however, the transfers are not budget balanced. When the agents adopt truthful strategies, the total amount of monetary transfers the network manager receives from the agents is negative. This means that the mechanism runs large deficits subsidizing agents. In this section we show that this budget deficit can be alleviated by introducing a set of participation fees.

At time $t = 0$ and before realizing the first period's security states θ_0^i, each agent i can decide whether or not to participate in the mechanism[1]. If he decides to participate, he should pay a participation fee $\tilde{p}_0^i$. We construct participation fees such that in expectation, their total amount is equal to the total amount of future subsidies. We define the participation fee of agent i by

$$\tilde{p}_0^i = \frac{-1}{N-1} \sum_{j \neq i} \mathbb{E}\{\sum_{t=0}^{\infty} \delta^t p_t^j\}, \quad (13)$$

---

[1] Equivalently, we can assume that all agents start from the safe state $\theta_0^i = 1$.

where the expectation is taken with respect to agents' strategies determined by the mechanism, the initial distribution of the security states which is assumed to be known to the network manager and the agents, and the dynamics of the security network. Adding these fees balances the budget as

$$\sum_i \tilde{p}_0^i + \mathbb{E}\{\sum_{t=0}^{\infty} \delta^t p_t^i\} = -\mathbb{E}\{\sum_{t=0}^{\infty} \delta^t p_t^i\} + \mathbb{E}\{\sum_{t=0}^{\infty} \delta^t p_t^i\} = 0. \tag{14}$$

Therefore, the DCI mechanism with participation fees is ex-ante budget balanced. With the introduction of the participation fees, an agent might rather stay out of the mechanism to avoid paying the participation fee while he still enjoys the positive externality that he receives from other agents' participation in the mechanism. Below, we show that for sufficiently patient agents, all agents voluntarily participate in the DCI mechanism with participation fees.

Theorem 2. For δ sufficiently close to 1, the DCI mechanism with participation fees is ex-ante budget balanced, satisfies the IC and IR constraints, and maximizes the social welfare W.

4 Dynamic Optimal Policy for the Network Manager

In this section, we study the control problem that the network manager must solve to find an optimal allocation policy π^*, when the agents reveal their security states $\{\boldsymbol{\theta}_t\}$ truthfully. In this case, the network manager is faced with a Markov decision process (MDP) with perfect observations, where the transition probabilities are given by (1) and (2) and the instantaneous reward is given by $r(\boldsymbol{\theta}_t, a_t) := \sum_{i=1}^{n} v^i(\boldsymbol{\theta}_t, a_t)$. Using dynamic programming [14], the network manager can solve this problem numerically, and find an optimal policy. However, there are some settings where qualitative properties of an optimal policy can be derived analytically. In the following, we discover qualitative properties of an optimal policy within the context of a specific network topology.

Example. Consider a circular network with $n = 4$ agents, where $h = 1$, $d_i = 0$, and $l_{ij} = l \leq 0.5$, for all i, j that are adjacent agents. The next proposition fully describes an optimal policy for this setting and the behavior of the corresponding value function.

Proposition. (i) An optimal policy π^* applies the security measure to one of the head ends of the shortest 'run of unsafe agents'. A run of unsafe agents of length k is a succession of k unsafe agents consecutively located between two safe agents.

(ii) The value function $V^*(.)$ induces a complete ordering on the set of states, such that a state with a greater number of safe agents is strictly preferred to a state with smaller number of safe agents. In the case of equality, the state with a longer run of unsafe agents is strictly preferred.

The above proposition provides two metrics in comparing security states: (1) the number of safe agents and (2) how close the unsafe agents are to one another. Numerical results show that these two metrics still work in symmetric circular networks with an arbitrary number of agents. This means that if l is below a certain threshold, an optimal policy tries to first maximize the number of safe agents, and then, bring the unsafe agents close to one another. To do so, the network manager applies the security measure to one of the head ends of the shortest run of unsafe agents.

References

1. Athey, S., Segal, I.: An efficient dynamic mechanism. Econometrica **81**(6), 2463–2485 (2013)
2. Bergemann, D., Välimäki, J.: Information acquisition and efficient mechanism design. Econometrica **70**(3), 1007–1033 (2002)
3. Börgers, T., Krahmer, D., Strausz, R.: An Introduction to the Theory of Mechanism Design. Oxford University Press, Oxford (2015)
4. Chen, L., Leneutre, J.: A game theoretical framework on intrusion detection in heterogeneous networks. IEEE Trans. Inf. Forensics Secur. **4**(2), 165–178 (2009)
5. Cremer, J., McLean, R.P.: Full extraction of the surplus in bayesian and dominant strategy auctions. Econometrica **56**, 1247–1258 (1988)
6. Farhadi, F., Golestani, S.J.: Mechanism design for network resource allocation: a surrogate optimization approach (2016). http://ee.sharif.edu/~farhadi/Surrogate-Optimization-Approach.pdf
7. Farhadi, F., Tavafoghi, H., Teneketzis, D., Golestani, S.J.: A dynamic incentive mechanism for security in networks of interdependent agents (2016). http://ee.sharif.edu/~farhadi/Dynamic-Incentive-Mechanism.pdf
8. Fazeli, A., Ajorlou, A., Jadbabaie, A.: Competitive diffusion in social networks: quality or seeding? IEEE Trans. Control Netw. Syst. **PP**(99), 1 (2016)
9. Jackson, M., Zenou, Y.: Games on networks. In: Handbook of Game Theory with Economic Applications, vol. 4. Elsevier (2015)
10. Jiang, L., Anantharam, V., Walrand, J.: How bad are selfish investments in network security? IEEE/ACM Trans. Netw. **19**(2), 549–560 (2011)
11. Kakhbod, A., Teneketzis, D.: An efficient game form for unicast service provisioning. IEEE Trans. Autom. Control **57**(2), 392–404 (2012)
12. Kelly, F., Maulloo, A., Tan, D.: Rate control for communication networks: shadow prices, proportional fairness and stability. J. Oper. Res. Soc. **49**, 237–252 (1998)
13. Khouzani, M.H.R., Sarkar, S., Altman, E.: A dynamic game solution to malware attack. In: IEEE INFOCOM, April 2011
14. Kumar, P.R., Varaiya, P.: Stochastic Systems: Estimation, Identification, and Adaptive Control, vol. 75. SIAM (2015)
15. Laszka, A., Felegyhazi, M., Buttyan, L.: A survey of interdependent information security games. ACM Comput. Surv. **47**(2), 1–38 (2014)
16. Lelarge, M., Bolot, J.: Economic incentives to increase security in the internet: the case for insurance. In: IEEE INFOCOM, pp. 1494–1502 (2009)
17. Li, M., Koutsopoulos, I., Poovendran, R.: Optimal jamming attacks and network defense policies in wireless sensor networks. In: IEEE INFOCOM, May 2007
18. Liang, X., Xiao, Y.: Game theory for network security. IEEE Commun. Surv. Tutor. **15**(1), 101–120 (2013)

19. Liu, H.: Efficient dynamic mechanisms in environments with interdependent valuations. SSRN 2504731 (2014)
20. Myerson, R.B.: Multistage games with communication. Econom. J. Econom. Soc. **54**(2), 323–358 (1986)
21. Naghizadeh, P., Liu, M.: On the role of public and private assessments in security information sharing agreements. arXiv preprint arXiv:1604.04871 (2016)
22. Naghizadeh, P., Liu, M.: Opting out of incentive mechanisms: a study of security as a non-excludable public good. IEEE Trans. Inf. Forensics Secur. **11**(12), 2790–2803 (2016)
23. Nowzari, C., Preciado, V.M., Pappas, G.J.: Analysis and control of epidemics: a survey of spreading processes on complex networks. arXiv:1505.00768 (2015)
24. Trajanovski, S., Hayel, Y., Altman, E., Wang, H., Mieghem, P.V.: Decentralized protection strategies against SIS epidemics in networks. IEEE Trans. Control Netw. Syst. **2**(4), 406–419 (2015)

Rules for Computing Resistance of Transitions of Learning Algorithms in Games

Mohammed Shabbir Ali[1]([✉]), Pierre Coucheney[2], and Marceau Coupechoux[1]

[1] LTCI, Telecom ParisTech, Université Paris-Saclay, Paris, France
mdshabbirali88@gmail.com, marceau.coupechoux@telecom-paristech.fr
[2] DAVID-Lab, UVSQ, Versailles, France
pierre.coucheney@uvsq.fr

Abstract. In a finite game the Stochastically Stable States (SSSs) of adaptive play are contained in the set of minimizers of resistance trees. Also, in potential games, the SSSs of the log-linear learning algorithm are the minimizers of the potential function. The SSSs can be characterized using the resistance trees of a Perturbed Markov Chain (PMC), they are the roots of minimum resistance tree. Therefore, computing the resistance of trees in PMC is important to analyze the SSSs of learning algorithms. A learning algorithm defines the Transition Probability Function (TPF) of the induced PMC on the action space of the game. Depending on the characteristics of the algorithm the TPF may become composite and intricate. Resistance computation of intricate functions is difficult and may even be infeasible. Moreover, there are no rules or tools available to simplify the resistance computations. In this paper, we propose novel rules that simplify the computation of resistance. We first, give a generalized definition of resistance that allows us to overcome the limitations of the existing definition. Then, using this new definition we develop the rules that reduce the resistance computation of composite TPF into resistance computation of simple functions. We illustrate their strength by efficiently computing the resistance in log-linear and payoff-based learning algorithms. They provide an efficient tool for characterizing SSSs of learning algorithms in finite games.

Keywords: Potential games · Learning algorithms · Log-linear learning · Perturbed Markov Chains · Resistance of transitions

1 Introduction

In a finite repeated game if players sometimes make mistakes in choosing an optimal strategy and if all mistakes are possible and are time-independent then a perturbed Markov process is induced on the action space of the game. As the probability of mistakes goes to zero the stationary distribution of the process concentrates on particular equilibria. These are known as stochastically stable equilibria or Stochastically Stable States (SSS) of the game [1]. The SSSs correspond to the roots of minimum resistance trees where the resistance of a transition in a tree can be seen as the cost of deviating from the optimal strategy [2].

© ICST Institute for Computer Sciences, Social Informatics and Telecommunications Engineering 2017
L. Duan et al. (Eds.): GameNets 2017, LNICST 212, pp. 97–107, 2017.
DOI: 10.1007/978-3-319-67540-4_9

Therefore, the computation of resistance of transitions of a Perturbed Markov Chain (PMC) is important.

The learning algorithm used by the player of the game defines the Transition Probability Function (TPF) of the induced PMC. Depending on the characteristics of the learning algorithm the TPF can be composite and intricate. The resistance computation of intricate TPF is difficult and may even be infeasible for some functions. Moreover, there are no rules and no tools available in the literature to simplify the computation of resistance. We focus on developing novel rules to simplify the resistance computations of a general class of TPF.

As the perturbation slowly decreases the limiting stationary distribution of a PMC exists and is unique [2]. The support of the stationary distribution is the root of the minimum resistance tree. Exploring these results many learning algorithms for games are analyzed in the literature. In the following, we discuss a few such algorithms.

A log-linear learning algorithm is used for a potential game that models the load balancing problem of a heterogeneous wireless network [3]. In this algorithm, the log of TPF is linear functions of the payoffs of the players [3–5]. This algorithm induces a PMC on the action space of the game. The convergence of this algorithms is analyzed as follows. First, using the TPF in (5) [5] the expression of resistance of transition is (6) [5] is obtained. We observe that the derivation of this expression requires a careful insight into the TPF to reduce it into a simplified form so that the resistance can be obtained. Otherwise, in case the TPT cannot be reduced into a simple form then the resistance may not be feasible to compute. Second, the resistance of a feasible path in a tree is obtained using the structure of potential games. Finally, the SSSs of the game are characterized by using the minimum resistance tree definition. A binary log-linear learning algorithm is a reduced information algorithm, in which the log TPF is a linear function of the two most recent payoffs [5]. This algorithm was used to distributively balance the loads in heterogeneous networks using near-potential games [6]. The computation resistance of transition is difficult in this case. The convergence of this algorithm to the SSSs of a potential game is analyzed in a similar way as in log-linear algorithm [5,6].

A payoff-based learning algorithm is obtained by combining log-linear algorithm and binary log-linear algorithm [5]. Due to the combination of two algorithms, the TPF is much involved. Therefore, the computation of resistance of transition is much involved and difficult. The convergence of this algorithm is also analyzed in a similar way as in log-linear algorithm. Adaptive play algorithm was applied to an acyclic game to characterize its SSSs using the resistance trees [2]. A class of trial and error learning algorithms for any finite game are also analyzed using the resistance trees [7,8]. Due to the different modes of learning in these algorithms, the TPF becomes complicated and the resistance computation is difficult.

In the above literature survey, we see that the computation of resistance is used for characterizing the SSSs of many learning algorithms in games. Therefore, in this paper, we develop new rules that ease the computation of resistance of intricate TPF. To do this, we first give a generalized definition of resistance for any positive function. The new definition overcomes the limitation of the existing old definition of resistance. For example, the limit in the old definition of resistance is not always feasible to evaluate for some functions, see Sect. 3. The new definition allows us to define resistance for any positive function. Thereby, allowing us to propose new rules for computing resistance. The proposed rules reduce the resistance computation of composite TPF into resistance computation of simple functions. These rules provide a powerful tool that can be used for analyzing the convergence properties of learning algorithms in finite games.

The rest of the paper is organized as follows. In Sect. 2, we give an overview of resistance trees of PMC. In Sect. 3, we present new rules for resistance computation and provide their proves. In Sect. 4, we illustrate the application of the proposed rules. Conclusions are summarized in Sect. 5.

2   Overview of Resistance Trees

In this section, we first give a brief overview of resistance trees of a PMC. Then, using resistance trees we illustrate the convergence of log-linear learning algorithm in potential games. For more details see [2,5].

2.1   Resistance Trees of PMC

A perturbed Markov process is characterized by a set $\{P^\tau\}$ of transition matrices over a state space X indexed by a parameter τ. Wherein, $\tau \in (0, \tau_h]$ is a parameter that controls the perturbation, τ_h is constant. Probabilities P_{ab}^0 and P_{ab}^τ denote the transition probabilities from state a to b in the unperturbed and the perturbed Markov chains, respectively. The definition of resistance of transitions and the definition of a regular perturbed Markov process are below [2].

Definition 1 (Resistance of transition). *A perturbed Markov process $\{P^\tau\}$ is a regular if it satisfies the following conditions [2]:*

1. *P^τ is aperiodic and irreducible for all $\tau \in (0, \tau_h]$,*
2. *$\lim_{\tau \to 0} P_{ab}^\tau = P_{ab}^0$,*
3. *for a strictly positive TPF P_{ab}^τ there exists a non-negative number R_{ab} called the resistance of transition such that $0 < \lim_{\tau \to 0^+} e^{\frac{R_{ab}}{\tau}} P_{ab}^\tau < \infty$.*

Note that if $P_{ab}^0 > 0$ then $R_{ab} = 0$.

A tree, T, rooted at a state a, is a set of $|X| - 1$ directed edges such that, from every other state a', there is a unique directed path in the tree to a. The resistance of the directed edge $a \to b$ is denoted as R_{ab}. The resistance of a rooted tree, T, is the sum of the resistances on its edges $R(T) = \sum_{a,b \in T} R_{ab}$. Let $\mathcal{T}(a)$ be defined as the set of trees rooted at the state a. The stochastic potential of the state a is defined as $\gamma(a) = \min_{T \in \mathcal{T}(a)} R(T)$. A minimum resistance tree is a tree that has the minimum stochastic potential, that is, any tree T that satisfies $R(T) = \min_{a \in X} \gamma(a)$.

The following theorem by [2, Lemma 1] gives the existence and uniqueness of the stationary distribution of a PMC.

Theorem 1. *Let* $\{P^\tau\}$ *be a regular perturbed Markov process, and for each* $\tau > 0$, *let* μ_τ *be the unique stationary distribution of* P^τ. *Then* $\lim_{\tau \to 0} \mu_\tau$ *exists and the limiting distribution* μ_0 *is a stationary distribution of* P^0. *The stochastically stable states are the roots of minimum resistance trees.*

2.2   Convergence of Log-Linear Learning Algorithm Using Resistance Trees [5]

Log-linear learning algorithm induces a regular perturbed Markov process over the action space X of a n-player potential game [5]. Let $a = (a_i, a_{-i})$ denotes an action profile of the players where a_i denote the action of player i and a_{-i} denotes the actions of all the other players. Let X_i and X_{-i} denote the action space of player i and action space of other players, respectively. Let $b = (a'_i, a_{-i})$ denotes another action profile where player i changes its action. For $a \in X$, let $\phi(a)$ and $U_i(a)$ denote the potential function and utility of player i, respectively. In a potential game, for all $a_i, a'_i \in X_i$ and for all $a_{-i} \in X_{-i}$, we have $\phi(a) - \phi(b) = U_i(a) - U_i(b)$. Assuming that the player is selected with uniform probability the transition probability function of log-linear learning algorithm is given as below [5, (5)].

$$P^\tau_{ab} = \frac{1}{n} \frac{\exp\left(\frac{U_i(a'_i, a_{-i})}{\tau}\right)}{\sum_{a_i \in X_i} \exp\left(\frac{U_i(a_i, a_{-i})}{\tau}\right)} \tag{1}$$

The first step in the proof of convergence is to derive an expression of resistance of transition. Let $V(a_{-i}) := \max_{a_i \in X_i} U_i(a_i, a_{-i})$ and $B_i(a_i)$ denotes the set of actions that have the maximum utility. Multiplying the numerator and denominator of (1) by $e^{\frac{V(a_{-i})}{\tau}}$, we obtain

$$P^\tau_{ab} = \frac{1}{n} \frac{\exp\left(\frac{V(a_{-i}) - U_i(a'_i, a_{-i})}{\tau}\right)}{\sum_{a_i \in X_i} \exp\left(\frac{V(a_{-i}) - U_i(a_i, a_{-i})}{\tau}\right)}. \tag{2}$$

After simplifying the above equation, we obtain

$$\lim_{\tau \to 0^+} \frac{P^\tau_{ab}}{\exp\left(\frac{V(a_{-i}) - U_i(a'_i, a_{-i})}{\tau}\right)} = \frac{1}{n \, |B_i(a_i)|}. \tag{3}$$

Since, the above limit is positive and finite the induced process is a regular Markov process and the resistance according to Definition 1 is

$$R_{ab} = V(a_{-i}) - U_i(a'_i, a_{-i}). \tag{4}$$

Second step is to obtain the resistance of a path in the resistance trees. This is obtained in Lemma [5, Lemma 3.2] that we present below.

Lemma 1. *Let* $\mathcal{P} = \{a^0 \to a^1 \to \dots \to a^m\}$ *and* $\mathcal{P}^R = \{a^m \to a^{m-1} \to \dots \to a^0\}$ *be feasible forward path and reverse path, respectively. If all the players* *in a n-player potential game with potential function* $\phi : X \to \mathcal{R}$, *adhere to log-linear learning algorithm then the difference of resistance of paths is*

$$R(\mathcal{P}) - R(\mathcal{P}^R) = \phi(a^0) - \phi(a^m). \tag{5}$$

The final step is to prove that the stochastically stable states of the log-linear algorithm are the potential function maximizers of the potential game. This is accomplished by using Lemma 1 and minimum resistance tree definition. The detailed proof of the following theorem can be found in Proposition [5, 3.1].

Theorem 2. *If all the players of a potential game adhere to log-linear learning algorithm then the stochastically stable states are the potential function maximizers.*

3   Rules for Computing Resistance

The resistance in Definition 1 can be computed in case the transition function can be factorized into a simple function and in case the limit can be evaluated as shown in Sect. 2.2. However, transition functions can be composite and intricate that cannot always be simplified. Moreover, the limit in Definition 1 cannot always be feasible to evaluate. For example, when $P_{ab}^\tau = \tau$, the limit cannot be evaluated. To overcome these limitations of Definition 1 we first give a new generalized definition of resistance that allows us to develop easy rules to compute the resistance of any positive function.

Let $o(.)$ and $\omega(.)$ denote little "o" order and little "ω" order, respectively.

Definition 2 (Resistance of positive function). *The resistance of a strictly positive function* $f(\tau)$ *is* $Res(f)$ *if there exists a strictly positive function* $g(\tau)$ *such that* $g \in o\left(e^{k/\tau}\right)$ *and* $g \in \omega\left(e^{-k/\tau}\right)$ *for any* $k > 0$; *and*

$$\lim_{\tau \to 0} \frac{f(\tau)}{g(\tau)e^{-\frac{Res(f)}{\tau}}} = 1. \tag{6}$$

Remark 1. Note that Definition 2 includes Definition 1, in which $g(\tau) = \kappa, 0 < \kappa < \infty$. Now, we can evaluate the resistance of $P_{ab}^\tau = \tau$, i.e., $Res(\tau) = 0$.

Remark 2. Note that (6) is equivalent to

$$f(\tau) = g(\tau)e^{-\frac{Res(f)}{\tau}} + h(\tau), \tag{7}$$

where $h(\tau) \in o\left(g(\tau)e^{-\frac{Res(f)}{\tau}}\right)$.

Remark 3. We call $g(\tau)$ as a *sub-exponential function* if $g \in o\left(e^{k/\tau}\right)$ and $g \in \omega\left(e^{-k/\tau}\right)$ for any $k > 0$. Note that it is equivalent to $|\log g| \in o\left(\frac{1}{\tau}\right)$.

Lemma 2. *Consider any two sub-exponential functions $g_1(\tau)$ and $g_2(\tau)$. Consider two real numbers R_1 and R_2. If $R_1 < R_2$ then*

$$g_2(\tau)e^{-R_2/\tau} \in o\left(g_1(\tau)e^{-R_1/\tau}\right). \tag{8}$$

Proof. Let k be a real number. Then

$$\lim_{\tau \to 0} \frac{g_2(\tau)e^{-R_2/\tau}}{g_1(\tau)e^{-R_1/\tau}} = \lim_{\tau \to 0} \frac{g_2(\tau)}{e^{-(R_2-k)/\tau}} \left[\frac{g_1(\tau)}{e^{-(R_1-k)/\tau}}\right]^{-1}. \tag{9}$$

The above limit goes to zero when we choose $R_1 < k < R_2$. This is because the first factor goes to zero as $R_2 - k > 0$. Also, the second factor goes to zero as $R_1 - k < 0$. Recall that it is because g_1 and g_2 are sub-exponential.

Lemma 3. *If $Res(f)$ exists then it is unique.*

Proof. Assume that function f have two different resistances R_1 and R_2. Then, there exist g_1, g_2, h_1, h_2 such that

$$f(\tau) = g_1(\tau)e^{-\frac{R_1}{\tau}} + h_1(\tau) = g_2(\tau)e^{-\frac{R_2}{\tau}} + h_2(\tau), \tag{10}$$

where $h_1(\tau) \in o\left(g_1(\tau)e^{-\frac{R_1}{\tau}}\right)$ and $h_2(\tau) \in o\left(g_2(\tau)e^{-\frac{R_2}{\tau}}\right)$. Let $R_1 < R_2$. Using Lemma 2, we have $h_2 \in o\left(g_1(\tau)e^{-\frac{R_1}{\tau}}\right)$. Rearranging terms in (10), we have

$$1 + \frac{h_1(\tau)}{g_1(\tau)e^{-\frac{R_1}{\tau}}} = \frac{g_2(\tau)e^{-\frac{R_2}{\tau}}}{g_1(\tau)e^{-\frac{R_1}{\tau}}} + \frac{h_2(\tau)}{g_1(\tau)e^{-\frac{R_1}{\tau}}}. \tag{11}$$

Using Lemma 2 to evaluate the limit of the above equation as τ goes to zero, we arrive at contradiction that $1 = 0$.

The following proposition gives the rules for computing $Res(f)$.

Proposition 1. *Let f, f_1 and f_2 be strictly positive functions. Let κ be a positive constant. If $Res(f_1)$ and $Res(f_2)$ exist then*

I *$f_1(\tau)$ is sub-exponential if and only if $Res(f_1) = 0$. In particular $Res(\kappa) = 0$,*
II *$Res(e^{-\kappa/\tau}) = \kappa$,*
III *$Res(f_1 + f_2) = \min\{Res(f_1), Res(f_2)\}$,*
IV *$Res(f_1 - f_2) = Res(f_1), if\, Res(f_1) < Res(f_2)$,*
V *$Res(f_1 f_2) = Res(f_1) + Res(f_2)$,*
VI *$Res(\frac{1}{f}) = -Res(f)$,*
VII *If $f_1(\tau) \le f_2(\tau)$, $Res(f_1)$ and $Res(f_2)$ exist then $Res(f_2) \le Res(f_1)$,*
VIII *Let $f_1(\tau) \le f(\tau) \le f_2(\tau)$, If $Res(f_1) = Res(f_2)$ then $Res(f)$ exists and $Res(f) = Res(f_1)$.*

Remark 4. In Rule IV, if $Res(f_1) = Res(f_2)$ then we cannot compute $Res(f_1 - f_2)$ because in general the difference of sub-exponential functions may not be a sub-exponential function. For example, choose $f_1(\tau) = 1 + e^{-k/\tau}$ and $f_2(\tau) = 1$ with $k > 0$ then $Res(f_1) = Res(f_2) = 0$ but $Res(f_1 - f_2) = k$.

Remark 5. For Rule VIII, in general if $f_1(\tau) \leq f(\tau) \leq f_2(\tau)$ and $\mathrm{Res}(f_1) \neq \mathrm{Res}(f_2)$ then $\mathrm{Res}(f)$ may not exist. For example, for $f(\tau) = \lambda(\tau)f_1 + (1-\lambda(\tau))f_2$, $\lambda(\tau) = \frac{1}{2}\left(\cos\left(\frac{1}{\tau}\right)+1\right)$ the $\mathrm{Res}(f)$ does not exist.

Proof. Proof of Rule I: Let $f(\tau)$ be a sub-exponential function. Choosing $g(\tau) = f(\tau)$ and substituting $\mathrm{Res}(f) = 0$ in (6) we get $\lim_{\tau \to 0} \frac{f(\tau)}{f(\tau)e^{-\frac{\mathrm{Res}(f)}{\tau}}} = 1$. Therefore, we have $\mathrm{Res}(f) = 0$.

Assume $\mathrm{Res}(f) = 0$. From (7), we have $f(\tau) = g(\tau) + h(\tau)$, which is a sub-exponential function.

Let $f(\tau) = \kappa$ and $g(\tau) = \kappa$ then $g(\tau) \in o\left(e^{\frac{\kappa}{\tau}}\right)$ and $g(\tau) \in \omega\left(e^{-\frac{\kappa}{\tau}}\right)$, $\kappa > 0$. Substituting these in (6) we have $\mathrm{Res}(\kappa) = 0$.

Proof of Rule II: Substituting $f(\tau) = e^{-\kappa/\tau}$ and $g(\tau) = 1$ in (6) we get $\mathrm{Res}(f) = \kappa$.

Proof of Rule III: Let $\mathrm{Res}(f_1)$ and $\mathrm{Res}(f_2)$ be the resistances of functions f_1 and f_2, respectively. Then, from (7) we have $f_1(\tau) = g_1(\tau)e^{-\frac{\mathrm{Res}(f_1)}{\tau}} + h_1(\tau)$, $f_2(\tau) = g_2(\tau)e^{-\frac{\mathrm{Res}(f_2)}{\tau}} + h_2(\tau)$, where $h_1(\tau) \in o\left(g_1(\tau)e^{-\frac{\mathrm{Res}(f_1)}{\tau}}\right)$, $h_2(\tau) \in o\left(g_2(\tau)e^{-\frac{\mathrm{Res}(f_2)}{\tau}}\right)$. The sum of two functions can be written as

$$f_1(\tau) + f_2(\tau) = g_1(\tau)e^{-\frac{\mathrm{Res}(f_1)}{\tau}}\left(1 + \frac{h_1(\tau)}{g_1(\tau)e^{-\frac{\mathrm{Res}(f_1)}{\tau}}}\right.$$
$$\left. + \frac{g_2(\tau)e^{-\frac{\mathrm{Res}(f_2)}{\tau}}}{g_1(\tau)e^{-\frac{\mathrm{Res}(f_1)}{\tau}}} + \frac{h_2(\tau)}{g_1(\tau)e^{-\frac{\mathrm{Res}(f_1)}{\tau}}}\right), \tag{12}$$

Consider the case when $\mathrm{Res}(f_1) < \mathrm{Res}(f_2)$. Using Lemma 2 we have $h_2 \in o\left(g_1(\tau)e^{-\frac{\mathrm{Res}(f_1)}{\tau}}\right)$. Therefore, $f_1(\tau) + f_2(\tau) = g_1(\tau)e^{-\frac{\mathrm{Res}(f_1)}{\tau}} + h_3(\tau)$, where $h_3(\tau) \in o\left(g_1(\tau)e^{-\frac{\mathrm{Res}(f_1)}{\tau}}\right)$. According to (7), we have $\mathrm{Res}(f_1 + f_2) = \mathrm{Res}(f_1)$.

The case of $\mathrm{Res}(f_1) = \mathrm{Res}(f_2)$ leads to the same result as shown below.

$$f_1(\tau) + f_2(\tau) = e^{-\frac{\mathrm{Res}(f_1)}{\tau}}[g_1(\tau) + g_2(\tau)] + h_1(\tau) + h_2(\tau). \tag{13}$$

Note that sum of sub-exponential functions $g_1(\tau) + g_2(\tau)$ is a sub-exponential function. Observe that $h_1(\tau) + h_2(\tau) \in o\left([g_1(\tau) + g_2(\tau)]e^{-\frac{\mathrm{Res}(f_1)}{\tau}}\right)$. As in the previous case, according to (7) we have $\mathrm{Res}(f_1 + f_2) = \mathrm{Res}(f_1)$

Proof of Rule IV: Also, it can be shown similarly to the proof of rule III that if $\mathrm{Res}(f_1) < \mathrm{Res}(f_2)$ then $\mathrm{Res}(f_1 - f_2) = \mathrm{Res}(f_1)$.

Proof of Rule V:

$$\lim_{\tau \to 0} \frac{f_1(\tau)}{g_1(\tau)e^{-\frac{\mathrm{Res}(f_1)}{\tau}}} \lim_{\tau \to 0} \frac{f_2(\tau)}{g_2(\tau)e^{-\frac{\mathrm{Res}(f_2)}{\tau}}} = \lim_{\tau \to 0} \frac{f_1(\tau)f_2(\tau)}{g_1(\tau)g_2(\tau)e^{-\frac{\mathrm{Res}(f_1)+\mathrm{Res}(f_1)}{\tau}}} = 1. \tag{14}$$

Therefore, $\mathrm{Res}(f_1 f_2) = \mathrm{Res}(f_1) + \mathrm{Res}(f_2)$.

Proof of Rule VI: Since $\text{Res}(f)$ exists, inverting both sides of (6), we have

$$\lim_{\tau \to 0} \frac{f(\tau)}{g(\tau)e^{-\frac{\text{Res}(f)}{\tau}}} = 1 = \lim_{\tau \to 0} \frac{\frac{1}{f(\tau)}}{\frac{1}{g(\tau)}e^{-\frac{-\text{Res}(f)}{\tau}}}. \tag{15}$$

Note that $\frac{1}{g(\tau)}$ is sub-exponential. Therefore, we have $\text{Res}(\frac{1}{f}) = -\text{Res}(f)$.

Proof of Rule VII: Assume that $\text{Res}(f_1) < \text{Res}(f_2)$. Using Lemma 2, we have $g_2(\tau)e^{-\text{Res}(f_2)/\tau} \in o\left(g_1(\tau)e^{-\text{Res}(f_1)/\tau}\right)$ and $h_2 \in o\left(g_1(\tau)e^{-\text{Res}(f_1)/\tau}\right)$.

$$f_1 \le f_2, \tag{16}$$

$$g_1(\tau)e^{-\text{Res}(f_1)/\tau} + h_1(\tau) \le g_2(\tau)e^{-\text{Res}(f_2)/\tau} + h_2(\tau), \tag{17}$$

$$1 + \frac{h_1(\tau)}{g_1(\tau)e^{-\text{Res}(f_1)/\tau}} \le \frac{g_2(\tau)e^{-\text{Res}(f_2)/\tau} + h_2(\tau)}{g_1(\tau)e^{-\text{Res}(f_1)/\tau}}. \tag{18}$$

As $\tau \to 0$, we arrive at a contradiction that $1 \le 0$. Therefore, $\text{Res}(f_1) \ge \text{Res}(f_2)$.

Proof of Rule VIII: We have $1 \le \frac{f(\tau)}{f_1(\tau)} \le \frac{f_2(\tau)}{f_1(\tau)}$ and $\text{Res}\left(\frac{f_2(\tau)}{f_1(\tau)}\right) = \text{Res}(f_2) - \text{Res}(f_1) = 0$. By Rule I $\frac{f_2(\tau)}{f_1(\tau)}$ is sub-exponential. This implies that $\frac{f(\tau)}{f_1(\tau)}$ is also sub-exponential. Therefore, there exists $g_{01}(\tau)$ such that

$$1 = \lim_{\tau \to 0} \frac{\frac{f(\tau)}{f_1(\tau)}}{g_{01}(\tau)} = \lim_{\tau \to 0} \frac{f(\tau)}{g_{01}(\tau)g_1(\tau)e^{-\frac{\text{Res}(f_1)}{\tau}}} \lim_{\tau \to 0} \frac{g_1(\tau)e^{-\frac{\text{Res}(f_1)}{\tau}}}{f_1(\tau)}, \tag{19}$$

$$= \lim_{\tau \to 0} \frac{f(\tau)}{g_{01}(\tau)g_1(\tau)e^{-\frac{\text{Res}(f_1)}{\tau}}}, \tag{20}$$

where the product $g_{01}(\tau)g_1(\tau)$ is also a sub-exponential function. Therefore, $\text{Res}(f)$ exists and $\text{Res}(f) = \text{Res}(f_1) = \text{Res}(f_2)$.

4   Application of Proposed Rules

In this section, we illustrate the application and robustness of the proposed rules for computing the resistance of composite TPFs.

4.1   Resistance of Log-Linear Learning Algorithm

By using Rule V and VI the resistance of $\text{Res}\,(P_{ab}^\tau)$ (1) is obtained as below.

$$\text{Res}\,(P_{ab}^\tau) = \text{Res}\left(\frac{1}{n}\right) + \text{Res}\left(e^{\frac{U_i(a_i', a_{-i})}{\tau}}\right) - \text{Res}\left(\sum_{a_i \in X_i} e^{\frac{U_i(a_i, a_{-i})}{\tau}}\right). \tag{21}$$

Applying the Rule III to the above equation, we have

$$\text{Res}\,(P_{ab}^\tau) = \text{Res}\left(\frac{1}{n}\right) + \text{Res}\left(e^{\frac{U_i(a_i', a_{-i})}{\tau}}\right) - \min_{a_i \in X_i} \text{Res}\left(e^{\frac{U_i(a_i, a_{-i})}{\tau}}\right). \tag{22}$$

Applying the Rule I and II, we get

$$\text{Res}\,(P_{ab}^\tau) = -U_i(a_i', a_{-i}) - \min_{a_i \in X_i} \left(-U_i(a_i, a_{-i})\right) = V\,(a_{-i}) - U_i(a_i', a_{-i}). \tag{23}$$

4.2   Resistance of Payoff-Based Learning Algorithm

In this subsection, we illustrate the application of the proposed rules by obtaining the expression of resistance payoff-based algorithm as in [5, Claim 6.1]. Let denotes two states of PMC of this algorithm as $z^1 := [a^0, a^1, x^1]$ and $z^2 := [a^1, a^2, x^2]$, where a^0, a^1, a^2 are action profiles and x^1, x^2 denotes the vectors representing whether the players have experimented or not, $x_i^1 = 0$ and $x_i^2 = 1$ represents that the player i had experimented. The transition probability function of Payoff-based algorithm is much involved as can be seen in [5, Claim 6.1].

$$
P^\tau_{z^1 \to z^2} = \left(\prod_{i:x_i^1=0,x_i^2=0} (1 - e^{-\frac{m}{\tau}}) \right) \left(\prod_{i:x_i^1=0,x_i^2=1} \frac{e^{-\frac{m}{\tau}}}{|X_i|} \right)
$$
$$
\left(\prod_{i:x_i^1=1,a_i^2=a_i^0} \frac{e^{\frac{U_i(a^0)}{\tau}}}{e^{\frac{U_i(a^0)}{\tau}} + e^{\frac{U_i(a^1)}{\tau}}} \right) \left(\prod_{i:x_i^1=1,a_i^2=a_i^1} \frac{e^{\frac{U_i(a^1)}{\tau}}}{e^{\frac{U_i(a^0)}{\tau}} + e^{\frac{U_i(a^1)}{\tau}}} \right) \quad (24)
$$

Using the Rule V, we have

$$
\mathrm{Res}\,(P^\tau_{z^1 \to z^2}) = \sum_{i:x_i^1=0,x_i^2=0} \mathrm{Res}\left(1 - e^{-\frac{m}{\tau}}\right) + \sum_{i:x_i^1=0,x_i^2=1} \mathrm{Res}\left(\frac{e^{-\frac{m}{\tau}}}{|X_i|}\right)
$$
$$
\sum_{i:x_i^1=1,a_i^2=a_i^0} \mathrm{Res}\left(\frac{e^{\frac{U_i(a^0)}{\tau}}}{e^{\frac{U_i(a^0)}{\tau}} + e^{\frac{U_i(a^1)}{\tau}}}\right) + \sum_{i:x_i^1=1,a_i^2=a_i^1} \mathrm{Res}\left(\frac{e^{\frac{U_i(a^1)}{\tau}}}{e^{\frac{U_i(a^0)}{\tau}} + e^{\frac{U_i(a^1)}{\tau}}}\right) \quad (25)
$$

Applying the Rules III, IV, V, and VI, we have

$$
\mathrm{Res}\,(P^\tau_{z^1 \to z^2}) = \sum_{i:x_i^1=0,x_i^2=0} \min\left\{\mathrm{Res}\,(1), \mathrm{Res}\left(e^{-\frac{m}{\tau}}\right)\right\}
$$
$$
+ \sum_{i:x_i^1=0,x_i^2=1} \left[\mathrm{Res}\left(e^{-\frac{m}{\tau}}\right) + \mathrm{Res}\left(\frac{1}{|X_i|}\right)\right]
$$
$$
+ \sum_{i:x_i^1=1,a_i^2=a_i^0} \left[\mathrm{Res}\left(e^{\frac{U_i(a^0)}{\tau}}\right) - \mathrm{Res}\left(e^{\frac{U_i(a^0)}{\tau}} + e^{\frac{U_i(a^1)}{\tau}}\right)\right]
$$
$$
+ \sum_{i:x_i^1=1,a_i^2=a_i^1} \left[\mathrm{Res}\left(e^{\frac{U_i(a^1)}{\tau}}\right) - \mathrm{Res}\left(e^{\frac{U_i(a^0)}{\tau}} + e^{\frac{U_i(a^1)}{\tau}}\right)\right]
$$
$$
(26)
$$

Simplifying further by applying the Rules I and II, we get

$$\text{Res}\left(P_{z1 \to z2}^{\tau}\right) = \sum_{i:x_i^1=0, x_i^2=0} \min\{0, m\} + \sum_{i:x_i^1=0, x_i^2=1} [m]$$

$$+ \sum_{i:x_i^1=1, a_i^2=a_i^0} \left[-U_i(a^0) - \min\left\{-U_i(a^0), -U_i(a^1)\right\}\right]$$

$$+ \sum_{i:x_i^1=1, a_i^2=a_i^1} \left[-U_i(a^1) - \min\left\{-U_i(a^0), -U_i(a^1)\right\}\right] \quad (27)$$

Let $V(a^0, a^1) = \max\left\{U_i(a^1), U_i(a^2)\right\}$, then we have

$$\text{Res}\left(P_{z1 \to z2}^{\tau}\right) = \sum_{i:x_i^1=0, x_i^2=1} m + \sum_{i:x_i^1=1, a_i^2=a_i^0} \left(V(a^0, a^1) - U_i(a^0)\right)$$

$$+ \sum_{i:x_i^1=1, a_i^2=a_i^1} \left(V(a^0, a^1) - U_i(a^1)\right) \quad (28)$$

The above obtained expression of resistance is same as in [5, (13)], verifying it.

5 Conclusion

Novel rules are proposed for computing the resistance of transition of a perturbed Markov chain. These rules reduce the computation of resistance of composite and intricate transition probability function into the computation of resistance of simple functions. These rules are simple and yet are powerful. The strength of these rules is illustrated by using them to calculate efficiently the resistance of transition of the well-known log-linear learning algorithm and the payoff-based learning algorithm. These calculations are verified by comparing the obtained expressions with that of in the literature. These rules provide an efficient tool that can be used to characterize the stochastically stable states of learning algorithms in finite games. We hope to apply these rules for analyzing new algorithms based on perturbed Markov chains as well as new game settings like potential games with noisy rewards [9].

References

1. Foster, D., Young, P.: Stochastic evolutionary game dynamics. Theor. Popul. Biol. **38**(2), 219–232 (1990)
2. Young, H.P.: The evolution of conventions. Econometrica **61**, 57–84 (1993). Jan
3. Ali, M.S., Coucheney, P., Coupechoux, M.: Load balancing in heterogeneous networks based on distributed learning in potential games. In: International Symposium on Modeling and Optimisation in Mobile, Ad Hoc, and Wireless Networks, pp. 371–378, May 2015
4. Blume, L.E.: The statistical mechanics of strategic interaction. Games Econ. Behav. **5**(3), 387–424 (1993)

5. Marden, J.R., Shamma, J.S.: Revisiting log-linear learning: asynchrony, completeness and a payoff-based implementation. Games Econ. Behav. **75**, 788–808 (2012)
6. Ali, M.S., Coucheney, P., Coupechoux, M.: Load balancing in heterogeneous networks based on distributed learning in near-potential games. IEEE Trans. Wirel. Commun. **15**(7), 5046–5059 (2016)
7. Young, H.P.: Learning by trial and error. Games Econ. Behav. **65**(2), 626–643 (2009)
8. Pradelski, B.S., Young, H.P.: Learning efficient Nash equilibria in distributed systems. Games Econ. Behav. **75**, 882–897 (2012). July
9. Leslie, D.S., Marden, J.R.: Equilibrium selection in potential games with noisy rewards. In: Proceedings of the IEEE Network Games, Control and Optimization (NetGCooP), pp. 1–4 (2011)

Optimal Control of Multi-strain Epidemic Processes in Complex Networks

Elena Gubar[1], Quanyan Zhu[2]([⊠]), and Vladislav Taynitskiy[1]

[1] Faculty of Applied Mathematics and Control Processes,
Saint Petersburg State University, Universitetskii Prospekt 35, Petergof,
Saint-Petersburg 198504, Russia
`e.gubar@spbu.ru`, `tainitsky@gmail.com`
[2] Department of Electrical and Computer Engineering, Polytechnic School
of Engineering, New York University, Brooklyn, USA
`quanyan.zhu@nyu.edu`

Abstract. The emergence of new diseases, such as HIV/AIDS, SARS, and Ebola, represent serious problems for the public health and medical science research to address. Despite the rapid development of vaccines and drugs, one challenge in disease control is the fact that one pathogen sometimes generates many strains with different spreading features. Hence it is of critical importance to investigate multi-strain epidemic dynamics and its associated epidemic control strategies. In this paper, we investigate two controlled multi-strain epidemic models for heterogeneous populations over a large complex network and obtain the structure of optimal control policies for both models. Numerical examples are used to corroborate the analytical results.

Keywords: Bi-virus models · Epidemic process · Optimal control · Structured population

1 Introduction

Infectious diseases remain a serious medical burden all around the world with 15 million deaths per year estimated to be directly related to infectious diseases. The emergence of new diseases such as HIV/AIDS, the severe acute respiratory syndrome (SARS) and, most recently, the rise of Ebola, represent a few examples of the serious problems that the public health and medical science research need to address.

While for centuries mankind seemed helpless against these sudden epidemics, in recent time, our ability to control future epidemic outbreaks has been facilitated by the advances in modern science. The cures for a number of dangerous pathogens are available and can be developed and manufactured faster than ever before thanks to the genetic revolution new drugs to prevent and reduce the health consequences of new epidemics. The vaccine against new influenza A (H1N1) has been developed rapidly to be available only a few months after the beginning of the epidemic.

© ICST Institute for Computer Sciences, Social Informatics and Telecommunications Engineering 2017
L. Duan et al. (Eds.): GameNets 2017, LNICST 212, pp. 108–117, 2017.
DOI: 10.1007/978-3-319-67540-4_10

However, one challenge in disease control is the fact that one pathogen sometimes generates many strains with different spreading features, and hence a detailed investigation of multi-strain epidemic dynamics has great relevance [1–3]. For example, the human immunodeficiency virus (HIV) (which causes acquired immune deficiency syndrome (AIDS)) has many genetic varieties and can be divided into several distinct strains, such as strain HIV-1 and strain HIV-2 [4]. On the other hand, one pathogen is always incorporated with other pathogens [5]. The influenza A (H1N1) virus has the potential to develop into the first influenza pandemic of the twenty-first century [6], and it is accompanied by seasonal influenza [7].

In this paper, we establish a control-theoretic model to design disease control strategies through quarantine and immunization to mitigate the impact of epidemics on our society. Disease transmission in epidemics can be represented by dynamics on a graph where vertices denote individuals and an edge connecting a pair of vertices indicates an interaction between individuals. Due to a large population of people involved in the process of disease transmission, random graph models such as the small-world networks in [8] or scale-free networks in [9] are convenient to capture the heterogeneous patterns in the large-scale complex network.

We investigate two controlled multi-strain epidemic models for heterogeneous populations over a large complex network. One is the Susceptible-Infected-Recovered (SIR) epidemic process. The control is to quarantine a fraction of the infected nodes. Another model is the Susceptible-Infected-Susceptible (SIS) epidemic process. The control in this model is to provide treatment to the infected individuals, while treated individuals can become susceptible again to the infection of the disease.

The paper is organized as follows. Section 2 presents the controlled SIR mathematical model. Section 3, using Pontryagin's minimum principle, defines the structure optimal control policies. Section 4 presents the optimal control problem for controlled SIS model. Section 5 focuses on the analysis of the optimal control of SIS model. Numerical examples will be presented in Sect. 6. Section 7 concludes the paper and presents future research directions.

2 SIR Model for Two-Strain Viruses

Denote by $S_k(t), R_k(t)$ the population densities of the *Susceptible* and *Recovered* nodes with degree k at time t. We consider two strains of viruses co-exist in the network. $I_k^1(t), I_k^2(t)$ are the population densities of the *Infected* nodes of degree k at time t. We assume that the total population is constant in the network for all t, i.e., $S_k(t) + I_k^1(t) + I_k^2(t) + R_k(t) = 1$. We have extended the simple SIR model introduced by [10] to describe the situation with two virus types over a complex network.

$$\begin{aligned}
\frac{dS_k}{dt} &= -\delta_1 S_k I_k^1 \Theta_1 - \delta_2 S_k I_k^2 \Theta_2; \\
\frac{dI_k^1}{dt} &= (\delta_1 S_k \Theta_1 - \sigma_1 - u_k^1) I_k^1; \\
\frac{dI_k^2}{dt} &= (\delta_2 S_k \Theta_2 - \sigma_2 - u_k^2) I_k^2; \\
\frac{dR_k}{dt} &= (\sigma_1 + u_k^1) I_k^1 + (\sigma_2 + u_k^2) I_k^2,
\end{aligned} \tag{1}$$

where δ_i are infection rates for virus V_i, $i = 1, 2$, and σ_i are recovered rates.

At the beginning of epidemic process $t = 0$, most of nodes in the network belong to the susceptible subgroup, and small subgroup in total population is infected; and the remaining nodes are in the recovered subgroup. Hence initial states are $0 < S_k(0) < 1$, $0 < I_k^1(0) < 1$, $0 < I_k^2(0) < 1$, $R_k(0) = 1 - S_k(0) - I_k^1(0) - I_k^2(0)$. $\Theta_i(t)$ can be written in general (see [11], [12]) as

$$\Theta_i(t) = \sum_{k'} \frac{\tau(k')P(k'|k)I_{k'}^j}{k'}, \quad i = 1, 2, \tag{2}$$

where $\tau(k)$ denotes the infectivity of a node with degree k. $P(k'|k)$ describes the probability of a node with degree k pointing to a node with degree k', and $P(k'|k) = \frac{k'P(k')}{\langle k \rangle}$, where $\langle k \rangle = \sum_{k'} kP(k)$. For scale-free node distribution $P(k) = C^{-1}k^{-2-\gamma}$, $0 < \gamma \le 1$, where $C = \zeta(2 + \gamma)$ is Riemann's zeta function, which provides an appropriate normalization constant for sufficiently large networks.

The control parameters which can be used to protect the network from the propagation of the virus with k links are defined as $u_k = (u_k^1, u_k^2)$. Here, u_k^i are the fractions of the infected nodes which are quarantined in the population. The rates σ_i are the coefficients of "self-recovery", which can be interpreted as the activity of stationary antivirus software or firewalls.

The objective function: We minimize the overall cost in time interval $[0, T]$. At any given t, the following costs $f_1(I_k^1(t)), f_2(I_k^2(t))$ are treatment costs; $g(R_k(t))$ is utility of having $R_k(t)$ fraction of nodes recovered at time t; $h_1(u_k^1(t)), h_2(u_k^2(t))$ are costs for using antivirus patches or quarantine that help to reduce epidemic spreading, $k_{I_k^1}$, $k_{I_k^2}$, k_R represent the cost and benefit for the infected and the recovered in the end of the epidemic, respectively. Here, functions $f_i(t)$ are non-decreasing and twice-differentiable, convex functions, with $f_i(0) = 0$, $f_i(I_k^i) > 0$ for $I_k^i > 0, i = 1, 2$; $g(R_k(t))$ is non-decreasing and differentiable, and $g(0) = 0$; $h_i(u_k^i(t))$ is a twice-differentiable and increasing function in $u_k^i(t)$ such that $h_i(0) = 0$, $h_i(u_k^i) > 0$, $i = 1, 2$ when $u_k^i > 0$.

The aggregated system cost is given by

$$J = \int_0^T f_1(I_k^1(t)) + f_2(I_k^2(t)) - g(R_k(t)) + h_1(u_k^1(t))$$

$$+ h_2(u_k^2(t))dt + k_{I_k^1}I_k^1(T) + k_{I_k^2}I_k^2(T) - k_{R_k}R_k(T) \tag{3}$$

and the optimal control problem is to minimize the cost, i.e., $\min_{\{u_k^1, u_k^2\}} J$. To simplify the analysis, we consider the case where $k_{I_k^1} = k_{I_k^2} = k_{R_k} = 0$.

Treatment or isolation can be considered as the control parameters that can reduce the fraction of infected nodes in network. We define $u_k = (u_k^1, u_k^2)$ as control variables with $0 \le u_k^1(t) \le 1$, $0 \le u_k^2(t) \le 1$, for all t.

3   Optimal Control of SIR Model

We use Pontryagin's minimum principle [13] to find the optimal solution $u_k(t) = (u_k^1(t), u_k^2(t))$ to the problem described above. Define the associated Hamiltonian

H and adjoint functions λ_{S_k}, $\lambda_{I_k^1}$, $\lambda_{I_k^2}$, λ_{R_k} as follows:

$$
\begin{aligned}
H = {} & f_1(I_k^1(t)) + f_2(I_k^2(t)) - g(R_k(t)) + h_1(u_k^1(t)) \\
& + h_2(u_k^2(t)) + (\lambda_{I_k^1}(t) - \lambda_{S_k}(t))\delta_1 S_k(t) I_k^1(t)\Theta_1(t) \\
& + (\lambda_{I_k^2}(t) - \lambda_{S_k}(t))\delta_2 S_k(t) I_k^2(t)\Theta_2(t) \\
& + (\lambda_{R_k}(t) - \lambda_{I_k^1}(t))\sigma_1 I_k^1(t) + (\lambda_{R_k}(t) - \lambda_{I_k^2}(t))\sigma_2 I_k^2(t) \\
& - (\lambda_{I_k^1}(t) - \lambda_{R_k}(t))I_k^1(t)u_k^1 - (\lambda_{I_k^2}(t) - \lambda_{R_k}(t))I_k^2(t)u_k^2(t).
\end{aligned}
\tag{4}
$$

Here, we have used the condition $R = 1 - S_k - I_k^1 - I_k^2$. We construct the associated adjoint system as follows:

$$
\begin{aligned}
\dot{\lambda}_S(t) &= -\frac{\partial H}{\partial S} = -(\lambda_{I_k^1} - \lambda_{S_k})\delta_1 I_k^1\Theta_1 - (\lambda_{I_k^2} - \lambda_{S_k})\delta_2 I_k^2\Theta_2; \\
\dot{\lambda}_{I_k^1}(t) &= -\frac{\partial H}{\partial I_k^1} = -f_1'(I_k^1) + (\lambda_{S_k} - \lambda_{I_k^1})\delta_1 S_k\Theta_1 \\
&\qquad\qquad\quad - (\lambda_{R_k} - \lambda_{I_k^1})\sigma_1 + (\lambda_{I_k^1} - \lambda_{R_k})u_k^1; \\
\dot{\lambda}_{I_k^2}(t) &= -\frac{\partial H}{\partial I_k^2} = -f_2'(I_k^2) + (\lambda_{S_k} - \lambda_{I_k^2})\delta_2 S_k\Theta_2 \\
&\qquad\qquad\quad - (\lambda_{R_k} - \lambda_{I_k^2})\sigma_1 + (\lambda_{I_k^2} - \lambda_{R_k})u_k^2; \\
\dot{\lambda}_{R_k}(t) &= -\frac{\partial H}{\partial R_k} = g'(R_k);
\end{aligned}
\tag{5}
$$

with the transversality conditions given by

$$
\lambda_{I_k^1}(T) = 0, \; \lambda_{I_k^2}(T) = 0, \; \lambda_{S_k}(T) = 0, \; \lambda_{R_k}(T) = 0.
\tag{6}
$$

According to Pontryagin's minimum principle [13], there exist continuous and piecewise continuously differentiable co-state functions λ_i that at every point $t \in [0, T]$ where u_k^1 and u_k^2 is continuous, satisfying (5) and (6). In addition, we have

$$
(u_k^1, u_k^2) \in \arg \min_{u_k^1, u_k^2 \in [0,1]} H(\overline{\lambda}, (S_k, I_k^1, I_k^2, R_k), (\underline{u_k^1}, \underline{u_k^2})),
\tag{7}
$$

where $\overline{\lambda} = (\lambda_{S_k}, \lambda_{I_k^1}, \lambda_{I_k^2}, \lambda_{R_k})$.

4   Structure of Optimal Control

Based on previous research, e.g., [13–15], in this section, we show that an optimal control $u_k(t) = (u_k^1(t), u_k^2(t))$ has the structure summarized in Proposition 1.

Proposition 1. *The following statements hold for the optimal control problem described in Sect. 2:*

- *If $h_i(\cdot)$ are concave, then there exist time moment t_1 $(0 < t_1 < T)$ such that:*

$$
u_k^i(t) = \begin{cases} 1, & \text{for } \phi_k^i < h_i(1), \quad 0 < t < t_1; \\ 0, & \text{for } \phi_k^i > h_i(1), \quad t_1 < t < T. \end{cases}
$$

- *If $h_i(\cdot)$ are strictly convex, then exists t_0, t_1 $(0 < t_0 < t_1 < T)$:*

$$
u_k^i(t) = \begin{cases} 0, & \phi_k^i \leq h_i'(0), \; i = 1, 2, & t \in (t_1; T]; \\ h'^{-1}(\phi_k^i), & h_i'(0) < \phi_k^i \leq h_i'(1), \; i = 1, 2, \; t \in (t_0; t_1]; \\ 1, & h_i'(1) < \phi_k^i, \; i = 1, 2, & t \in [0; t_1]. \end{cases}
$$

Lemma 1. *Functions ϕ_k^i, $i = 1,2$ are decreasing functions of t, for all $t \in [0,T]$.*

Lemma 2. *For all $0 \leq t \leq T$, we have $(\lambda_{I_k^1} - \lambda_{S_k}) > 0$, $(\lambda_{I_k^2} - \lambda_{S_k}) > 0$, $(\lambda_{R_k} - \lambda_{I_k^1}) > 0$.*

The construction of optimal controls for the structured population follows the Pontryagin's minimum principle [13] and similar approaches used in [14], [15].

5   SIS Model with Two Virus Strains

A set of nodes N is divided into two subgroups: the Susceptible (S), the Infected (I). We suppose that two different viruses with different strains circulate in the network at time t. Let $S_k(t)$, $I_k^1(t)$, $I_k^2(t)$ be the densities of the susceptible and infected nodes with degree k at time t. $\lambda_i = \dfrac{\delta_i}{\sigma_i}$, where δ_i is infection rate and infected nodes are cured and become again susceptible with rate σ_i, $i = 1,2$.

$$
\begin{aligned}
\frac{dS_k}{dt} &= -\lambda_1 k S_k(t)\Theta_1 - \lambda_2 k S_k(t)\Theta_2 \\
&\quad + u_k^1 I_k^1(t) + u_k^2 I_k^2(t) + I_k^1(t) + I_k^2(t); \\
\frac{dI_k^1}{dt} &= \lambda_1 k S_k(t)\Theta_1 - I_k^1(t) - u_k^1(t) I_k^1(t); \\
\frac{dI_k^2}{dt} &= \lambda_2 k S_k(t)\Theta_2 - I_k^2(t) - u_k^2(t) I_k^2(t).
\end{aligned}
\tag{8}
$$

Objective function. We will minimize the overall cost in time interval $[0,T]$. At any given t, the following costs exist in the system: $f_i(I_k^i(t))$ are infected costs; $h_i(u_k^i(t))$ are costs for medical measures (i.e. quarantining) that help reduce the epidemic spreading. Here, the functions $f_i(I_k^i(t))$ are non-decreasing, twice-differentiable, and convex with $f_i(0) = 0$, $f_i(I_k^i(t)) > 0$ for $I_k^i > 0$, $g(S_k(t))$ non-decreasing and differentiable function, describing the benefits of using control, where $S_k(t) = 1 - I_k^1(t) - I_k^2(t)$ and $g(0) = 0$; $h_i(u_k^i(t))$ are twice-differentiable and increasing function in $u_k^i(t)$ such that $h_i(0) = 0$, $h_i(u_k^i) > 0$ when $u_k^i > 0$ with feasible controls $u_k^i \in [0,1]$.

The aggregated system cost is given by

$$
J = \int_0^T f_1(I_k^1(t)) + f_2(I_k^2(t)) + h_1(u_k^1(t))
$$

$$
+ h_2(u_k^2(t)) - g(S_k(t)) dt.
\tag{9}
$$

and the optimal control problem is to minimize the cost, i.e., $\min_{u_k^1, u_k^2 \in [0,1]} J$. System (8) describes the propagation of two different strains of viruses in the network. The propagation of the viruses is controlled by parameters u_k^i, $i = 1,2$. Here, u_k^i are antivirus policies.

We use Pontryagin's minimum principle to find the optimal control $u_k(t) = (u_k^1(t), u_k^2(t))$ which yields the minimum solution to the functional (9) for the problem described above. Consider the Hamiltonian:

$$
\begin{aligned}
H = &-l_0(f_1(I_k^1(t)) + f_2(I_k^2(t)) + h_1(u_k^1(t)) + h_2(u_k^2(t)) \\
&-g(S_k(t))) + l_1(t)(-\lambda_1(t)kS_k(t)\Theta_1(t) \\
&-\lambda_2(t)kS_k(t)\Theta_2(t) + u_k^1(t)I_k^1(t) + u_k^2(t)I_k^2(t) + I_k^1(t) \\
&+I_k^2(t)) + l_2(t)(\lambda_1 kS(t)\Theta_1(t) - I_k^1(t) - u_k^1(t)I_k^1(t)) \\
&+l_3(t)(\lambda_2(t)kS_k(t)\Theta_2(t) - I_k^2(t) - u_k^2(t)I_k^2(t)).
\end{aligned} \tag{10}
$$

where $l_0 = 1$. The adjoint systems are

$$
\begin{aligned}
\dot{l}_1(t) &= -\frac{\partial H}{\partial S_k} = -g'(S_k) - l_1(-\lambda_1\Theta_1 I_k^1 - l_2\lambda_2\Theta_2 I_k^2) - l_2\lambda_1\Theta_1 I_k^1 - l_3\lambda_2\Theta_2 I_k^2; \\
\dot{l}_2(t) &= -\frac{\partial H}{\partial I_k^1} = f_1'(I_k^1) - l_1(-\lambda_1\Theta_1 S_k + u_k^1 + 1) - l_2(\lambda_1 S_k\Theta_1 - 1 - u_k^1); \\
\dot{l}_3(t) &= -\frac{\partial H}{\partial I_k^2} = f_2'(I_k^2) - l_1(-\lambda_2 S_k\Theta_2 + u_k^2 + 1) - l_3(\lambda_2 S_k\Theta_2 - 1 - u_k^2),
\end{aligned} \tag{11}
$$

with the transversality condition:

$$
l_i(T) = 0. \tag{12}
$$

Consider next derivatives:

$$
\frac{\partial H}{\partial u_k^1} = h_1'(u_k^1) + (l_1 - l_2)I_k^1; \quad \frac{\partial H}{\partial u_k^2} = h_2'(u_k^2) + (l_1 - l_3)I_k^2. \tag{13}
$$

According to Pontryagin's minimum principle, there exist continuous and piecewise continuously differentiable co-state functions l_i that at every point $t \in [0, T]$ where u_k is continuous, satisfy (11) and (12). In addition, we have $l(t) = (l_0(t), l_1(t), l_2(t), l_3(t))$

$$
u_k^i \in \arg\max_{\underline{u}_k^i \in [0,1]} H(l, (S_k, I_k^1, I_k^2), \underline{u}_k^i). \tag{14}
$$

Since $h_i(u_k^i)$ is non-increasing function, then $h_i'(u_k^i) \geq 0$, $I_k^i \geq 0$ as a fraction of infected nodes, then condition (13) is satisfied only if $\psi_k^i > 0$, where

$$
\psi_k^1 = (l_1 - l_2)I_k^1; \quad \psi_k^2 = (l_1 - l_3)I_k^2. \tag{15}
$$

is defined as the switching function.

Then, to establish the optimal vaccination policy, we formulate the next proposition.

Proposition 2. *The optimal vaccination policy has following structure: If $h(\cdot)$ are concave, then exists time moment $0 < t_1 < T$ such that:*

$$
u_k^i(t) = \begin{cases} 0, & \text{if } \psi_k^i < h_i(1), \quad t \in (t_1; T]; \\ 1, & \text{if } \psi_k^i > h_i(1), \quad t \in [0; t_1]. \end{cases} \tag{16}
$$

If $h(\cdot)$ is strictly convex, then exists t_0, t_1 $(0 < t_0 < t_1 < T)$ such that:

$$
u_k^i(t) =
\begin{cases}
0 , & \text{if } \psi_k^i \leq \frac{\partial h_i(0)}{\partial u_k^i}, & t \in (t_1; T]; \\
h'^{-1}(\psi_k^i) , & \text{if } \frac{\partial h_i(0)}{\partial u_k^i} < \psi_k^i \leq \frac{\partial h_i(1)}{\partial u_k^i}, & t \in (t_0; t_1]; \\
1 , & \text{if } \psi_k^i > \frac{\partial h_i(1)}{\partial u_k^i}, & t \in [0; t_0].
\end{cases}
\tag{17}
$$

Lemma 3. *Functions $\dot{\psi}_i \leq 0$ are decreasing over the time interval $[0, T)$.*

Lemma 4. *Function $(l_1 - l_2) \leq 0$ and $(l_1 - l_3) \leq 0$ over the time interval $[0, T)$.*

To prove the proposition 2, we follow the same techniques as in [13], [14], [15].

6   Numerical Simulation

In this section, we present numerical simulations which are used to corroborate the results of main propositions. We depict optimal policies for SIR and SIS models for different cases if cost functions $h_i(u_k^i)$ are strictly convex and concave.

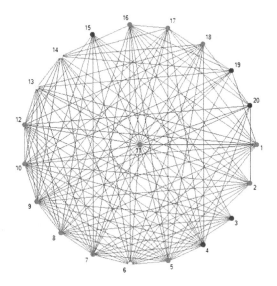

Fig. 1. The example of scale-free network for $N = 20$. Group $S = 5$ (blue dots), group $I = 3$, (yellow dots), group $R = 12$, (red dots). (Color figure online)

Here we take a piecewise linear infectivity,

$$
\tau(k) = \min(\alpha k, A),
\tag{18}
$$

where α and A are positive constants, $0 < \alpha \leq 1$. We set the infectivity parameter $\alpha = 0.02$, then for $k \in [1, 10]$ the infectivity rises and from $k > 10$ the individuals have the same infectivity equals to $A = 0.2$. To generate a net-work with scale-free exponent 3 we use the preferential attachment algorithm of Barabási and Albert (parameter $\gamma = 1$) [11,16]. The example of scale-free network for $\gamma = 1$, $N = 20$, $\langle k \rangle = 13.9$, maximum degree $k = 19$ is presented in Fig. 1 [17] (Figs. 2 and 3).

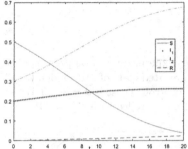

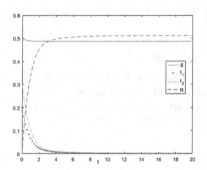

Fig. 2. Experiment I. SIR model with-out applying of control (degree $k = 10$). Initial states are $I_k^1(0) = 0.2$, $I_k^2(0) = 0.3$, the maximum values are $I_{1max} = 0.26$, $I_{2max} = 0.67$. Epidemic peaks are reached at $T = 20$. Average connectiv-ity $\langle k \rangle = 13.9$.

Fig. 3. Experiment I. SIR multi-strain controlled model (degree $k = 10$). Cost functions $h_i(\cdot)$ are strictly convex. Average connectivity $\langle k \rangle = 13.9$.

Experiment I. We use the following values for SIR model:initial fractions of susceptible, infected and recovered nodes are $S(0) = 0.5$, $I^1(0) = 0.2$, $I^2(0) = 0.3$ and $R(0) = 0$; infection rates are $\delta_1 = 0.3$ and $\delta_2 = 0.4$; recovered rates are $\sigma_1 = 0.003$ and $\sigma_2 = 0.001$; epidemic duration is $T = 20$ and costs function $f_{I_k^1} = 8I_k^1$, $f_{I_k^2} = 10I_k^2$, $g(R_k) = 0.1R_k$; $h_i(u_k^i)$ are convex functions $h_1(u_k^1) = 0.4(u_k^1)^2$ and $h_2(u_k^2) = 0.5(u_k^2)^2$. The optimal control policy is shown in Fig. 4.

Experiment II. Numerical simulations for SIS multi-strain model use the fol-lowing values: initial fractions of susceptible and infected nodes are $S(0) = 0.7$, $I^1(0) = 0.1$, $I^2(0) = 0.2$; infection rates are $\delta_1 = 0.3$ and $\delta_2 = 0.4$; recovered rates are $\sigma_1 = 0.003$ and $\sigma_2 = 0.001$; epidemic duration is $T = 20$ and costs func-tion $f_{I_k^1} = 8I_k^1$, $f_{I_k^2} = 10I_k^2$, $g(R_k) = 0.1R_k$; $h_i(u_k^i)$ are strictly convex functions $h_1(u_k^1) = 0.4(u_k^1)^2$ and $h_2(u_k^2) = 0.5(u_k^2)^2$ (Figs. 5, 6 and 7).

For both experiments, we have that the shape of control curves is the same for each k and we have used the same class of functionals for SIR and SIS dynamic systems.

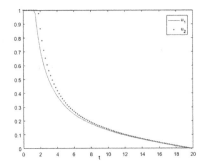

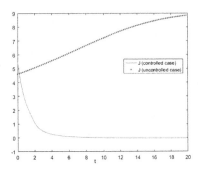

Fig. 4. Experiment I: Optimal control in SIR model, costs functions $h_i(u_k^i)$ are strictly convex. Switching points are $t_1 = 1.4$ and $t_2 = 1.8$.

Fig. 5. Experiment I: Comparison of the aggregated costs of SIR model: the cost of controlled case is $J = 36.39$, the cost of uncontrolled case is $J = 701.3$.

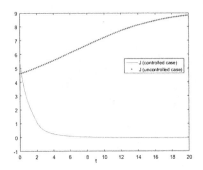

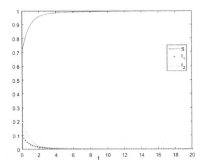

Fig. 6. Experiment II: SIS multi-strain model without control (degree $k = 10$). Initial states are $I_k^1(0) = 0.1$, $I_k^2(0) = 0.2$, the maximum values are $I_{1max} = 0.12$, $I_{2max} = 0.73$. Epidemic peaks are reached at $T = 20$. Average connectivity $\langle k \rangle = 13.9$.

Fig. 7. Experiment II: SIS multi-strain controlled model (degree $k = 10$). Cost functions $h_i(\cdot)$ are strictly convex. The average connectivity is $\langle k \rangle = 13.9$.

7   Conclusion

This paper has investigated the optimal control of two epidemic models of two co-existing virus strains for heterogeneous populations over a large complex network. We have obtained the structure of the optimal controller in the form of a threshold policy for a specific class of cost functions. Numerical examples have been used to corroborate the results. We would further explore the stability properties of the epidemic process under the optimal control.

Acknowledgement. The work of the second author is partially supported by the grant CNS-1544782, EFRI-1441140 and SES-1541164 from National Science Foundation (NSF).

References

1. Masuda, N., Konno, N.: Multi-state epidemic processes on complex networks. J. Theor. Biol. **243**(1), 64–75 (2006)
2. Nuno, M., Feng, Z., Martcheva, M., Castillo-Chavez, C.: Dynamics of two-strain influenza with isolation and partial cross-immunity. SIAM J. App. Math. **65**(3), 964–982 (2005)
3. Thomasey, D.H., Martcheva, M.: Serotype replacement of vertically transmitted diseases through perfect vaccination. J. Biol. Sys. **16**(02), 255–277 (2008)
4. Balter, M.: New HIV strain could pose health threat. Science **281**(5382), 1425–1426 (1998)
5. Castillo-Chavez, C., Huang, W., Li, J.: Competitive exclusion in gonorrhea models and other sexually transmitted diseases. SIAM J. App. Math. **56**(2), 494–508 (1996)
6. Smith, G.J.D., Vijaykrishna, D., Bahl, J., Lycett, S.J., Worobey, M., Pybus, O.G., Ma, S.K., Cheung, C.L., Raghwani, J., Bhatt, S.: Origins and evolutionary genomics of the 2009 swine-origin H1N1 influenza A epidemic. Nature **459**(7250), 1122–1125 (2009)
7. Butler, D.: Flu surveillance lacking. Nature **483**(7391), 520–522 (2012)
8. Strogatz, S.H.: Exploring complex networks. Nature **410**(6825), 268–276 (2001)
9. Barabási, A.-L., Réka, A.: Emergence of scaling in random networks. Science **286**(5439), 509–512 (1999)
10. Kermack, W.O., McKendrick, A.G.: A contribution to the mathematical theory of epidemics. In: Proceedings of the Royal Society of London A, vol. 115, issue 772, pp. 700–721. The Royal Society, New York (1927)
11. Fu, X., Small, M., Walker, D.M., Zhang, H.: Epidemic dynamics on scale-free networks with piecewise linear infectivity and immunization. Phys. Rev. E. **77**(3), 036113 (2008)
12. Pastor-Satorras, R., Vespignani, A.: Epidemic spreading in scale-free networks. Phys. Rev. Lett. **86**(14), 3200 (2001)
13. Pontryagin, L.S.: Mathematical Theory of Optimal Processes. CRC Press, Boca Raton (1987)
14. Khouzani, M.H.R., Sarkar, S., Altman, E.: Optimal control of epidemic evolution. In: INFOCOM IEEE Proceedings, pp. 1683–1691 (2011)
15. Gubar, E., Zhu, Q.: Optimal control of influenza epidemic model with virus mutations. In: Proceeding of European Control Conference (ECC), pp. 3125–3130 (2013)
16. Barabási, A.-L., Albert, R.: Statistical mechanics of complex networks. Rev. Mod. Phys. **74**, 47 (2002)
17. Gubar, E., Kumacheva, S., Zhitkova, E., Porokhnyavaya, O.: Impact of propagation information in the model of tax audit. In: Petrosyan, L.A., Mazalov, V.V. (eds.) Recent Advances in Game Theory and Applications, Static & Dynamic Game Theory: Foundations & Applications, pp. 91–110. Springer, Heidelberg (2016). doi:10.1007/978-3-319-43838-2_5

Invited Papers

Better Late Than Never: Efficient Transmission of Wide Area Measurements in Smart Grids

Jingchao Bao[⊠] and Husheng Li

University of Tennessee, Knoxville, TN 37996, USA
jcbao@utk.edu, husheng@eecs.utk.edu

Abstract. The persistent pursuit of reliability dates back to the birth of power system. In the era of smart grid, the harsh requirement extends to the whole system including communication infrastructure. The concentration of wide area synchronized measurements within large system is challenging. In this paper, we investigate the data aggregation issue of phasor measurement units (PMU) data stream in the synchrophasor network, where large latencies lead to unnecessary packet loss. We reduce the final packet loss rate by formulating the data aggregation problem as a multiple stopping time problem. Based on simulation, the success rate booms when compared with single optimal stopping time and multiple fixed-stopping time approaches. Our result could benefit the future development of protocol design, system state estimation and missing data recovery techniques.

1 Introduction

Nowadays, the synchronized PMUs based wide area measurement system (WAMS) accelerates the implementation of smart grid. Unlike the traditional power system, where measurements were gathered in supervisory control and data acquisition (SCADA) in an asynchronous fashion, all these synchronized measurements are marked with GPS time stamp and exchanged through the communication networks in real time to monitor, protect and control the dynamic operation of large area power system. The salient advantages of such system are the inborn time alignment and direct measurement of state instead of indirect system state estimation in the old time. The communication network becomes a critical component to build on. Just like a clot in the vein could cause severe damage to human brain, packet loss in a switch-based communication network for the smart grid system could blind the SCADA and lead to the catastrophic disasters.

However, most available communication infrastructures are built on the principle of probability and only promise to do the best under most circumstances. With no guarantee of the worst case packet loss and latency, the power community is reluctant to accept additional communication infrastructure although the potential benefits are huge. To reassure the doubt, more efforts should be made to mitigate and improve the system design considering the protocol, device and algorithm as a whole system.

© ICST Institute for Computer Sciences, Social Informatics and Telecommunications Engineering 2017
L. Duan et al. (Eds.): GameNets 2017, LNICST 212, pp. 121–130, 2017.
DOI: 10.1007/978-3-319-67540-4_11

The concept of phasor network was proposed by 1990s and has been implemented in the power system since the 21 century [1]. Within the rapid development of computing capability of modern computer and the boom of communication bandwidth in the past two decades, the once reasonable system design is worthy more consideration with new technology. Recently there are some debates on the existence of phasor data concentrator (PDC), mainly because of the additional latency introduced during the data transmission. Moreover, it appears that the time alignment function will magnify the packet loss problem in the large geographic system in two perspectives. The first factor is that current protocol will consider the messages arriving later than the deadline as lost, which converts part of the arrived packets as lost. The second negative factor is caused by the aggregated function in PDC. From the perspective of upper level receiver, a single packet loss from PDC means all the aggregated measurements from lower level are lost.

In [2], authors studied the missing data recovery using the matrix completion. However, it cannot recover the spike signal in the missing data. It is always better to attain the original measurements as much as possible when compared with possible post-recovery process. The authors in [3] discussed two scenarios where a dynamic waiting time is determined by the distributional information of all the latencies of different links. Then it becomes an optimal stopping time problem which could be solved by mathematical tools.

Unlike smart meter system, the phaser network are mostly constructed with wired network, especially fiber communication infrastructure. With the decreasing cost of bandwidth, it is preferable to trade bandwidth with reliability. Here we extend it to a multiple time data aggregation problem in one period with two simple observations. First, the communication bandwidth is considerably cheap compared with old days. We could watch 4K videos stream on-line while the sample of PMUs are on the level of kbps. Today most of the synchrophasor networks are connected with optical fiber, where the bandwidth could be considered as huge pipe carrying a small stream, yet the reliability are not full optimized. Second, a second chance for packets arriving later than a conservative deadline will always improve the system packet loss rate. There are physical laws which we cannot break in any situation. However, current PDCs may be conservative to limit the deadline to be far ahead of this physical limit. More aggregated packets consisting of later arrived measurements will provide a more comprehensive vision for SCADA. The simulation result validates our assumption, which is given later in details.

The remainder of this paper is organized as follows. The system structure of synchrophasor network and the details of multiple aggregation in PDC are briefed in Sect. 2. Then in Sect. 3 we will give our system model and algorithm. Later we analyze the performance bound on our algorithm. Numerical simulation is compared with the original one in Sect. 4, and conclusions are drawn in Sect. 5.

2   Background of Synchrophasor Network and System Structure

Based on current standards of synchrophasor [4–6], the synchrophasor network consists of PMUs and PDCs in which data streams initiate from the lower level substations where PMUs are located, and then are sent to PDCs in a real time fashion. Typically, one PDC could aggregate these data streams from multiple PMUs in various key locations. Then these intermediate nodes could implement various sophisticated functions on the data streams for monitoring, control or protection.

The ownership of these PDCs usually belongs to different utilities or ISOs. Therefore, the network topology could be complicated as a directed acyclic graph or simple as a tree where measurements are gathered in one or multiple place. The system structure is shown in Fig. 1.

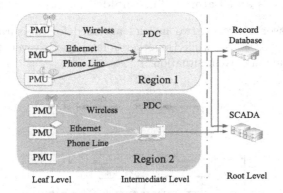

Fig. 1. Synchrophasor network

2.1   Data Aggregation in PDC

In the protocol [6], a PDC could perform different functions, such as data aggregation, data forwarding, data transfer protocols conversion, data latency calculation, redundant data handling etc., to relieve the burden of pre-process in the upper level control center. The specific configuration could be adjusted based on the need. This hierarchical structure usually offers great flexibility and scalability for a large distributed system.

Among all the functions, data aggregation and data forwarding are the most basic and core functions that PDC has. For data aggregation, it could be performed with or without time alignment. PDC should preserve data quality, time quality and time synchronization indication from each signal. For the case with time alignment, it refers to waiting for data with a given time stamp from all sources, placing that data in a packet, and forwarding it to next level. All the

data coming to a PDC has been timestamped by the PMU with a time refer-
enced to an absolute time. The PDC aligns received PMU/PDC data according
to their timestamps, not their arrival order or arrival time, and transmits the
combined data in one or more output data streams to other PDCs or applications
such as archiving, visualization, or control.

However, unlike most traffic in commercial networks, the measurement
streams in synchrophasor network is more time-critical. A large latency in
switched network reduces the value of measurements for some time-stringent
applications, especially protection or advanced control in the future. With data
aggregation enabled, each level in phasor network would set a latency deadline,
which inevitably introduces more latencies for the measurements arriving earlier
than the deadline. Moreover, it will further lessen the time conservation of data
transmission between this node and next level nodes, and the packets are prone
to losses. We will provide detailed discussions in the following.

2.2   Packet Loss and Latency Thresholds

For synchrophasor networks, two metrics are used for measuring the performance. One is packet loss, while the other is latency. However, the problem is
more troublesome in current situation.

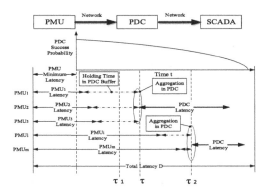

Fig. 2. Single deadline vs multiple grouping in PDC

Today, an empirical and conservative method is single fixed deadline policy,
where the waiting time is determined by the empirical latency measurements
and dedicated tuned with the best guess. The source of latency varies. To better
demonstrate the problem, we compare two scenarios in Fig. 2. Here we denote
the single waiting time by τ. With one time slot, there are m PMUs reporting
to a common PDC, whose latency is noted by L_i. The total latency allowance
D from PMUs to SCADA is determined by the specific application. The PDC
will wait and aggregate whatever it receives before the deadline τ and then
report to the SCADA in an integrated packet. From the view of SCADA, all

these packets have a latency equaling τ instead of $L_i \leq \tau_1$. On the other hand, it further compresses the transmission time for PDC from $D - L_i$ to $D - T_D$, which translates to a higher packet loss rate since it increases the probability that this integrated packet arrives later than D. In addition to that, these packets arrived after τ are ignored by the PDC and considered as lost from the view of SCADA.

Based on the above discussion, the latency in communication system exacerbates packet loss. With the advancement of communication techniques, the cost of reasonable high bandwidth decreases dramatically yet quality of service(QoS) such as packet loss rate has not been improved equally. The idea comes naturally to trade bandwidth for better packet loss rate. Specifically, it means we could arrange PDC to send multiple combined packets to SCADA instead a single one. The benefits are manifold. The first group of measurement sent by $\tau_1 \leq \tau$ has a better chance to reach SCADA in time. On the other hand, the next few groups, such as τ_2, are not abandoned from PDC and could have a considerate probability to be successfully received by the destination. The last but not the least, the shooting time for single deadline approach is really conservative since you do not want to take a risk to choose shooting time with a very low success rate. However, you can choose a shooting time in a larger time period than the single deadline approach, because it could be accepted in the multiple-time aggregation framework.

With all the benefits mentioned, how to choose the shooting time delicately to maximize the benefits remains a unsolved problem. The balance between shooting times and bandwidth cost has not been studied before in the literature. We will attack the problem in the next section by modeling it as a multiple optimal stopping time issue.

3   Multiple Optimal Stopping Time Problem in Data Aggregation

The requirement of latency, denoted by D, varies based the time scale of applications. However, some applications could be more stringent than others. The communication system should and has to satisfy the most stringent application with top priority. Here we assume that D is pre-determined.

3.1   Problem Formulation

Without loss of generality, we assume that one PDC has m PMU data streams to gather. Each link has latency L_i while the latency from PDC to control center is $\hat{L}$. In our analysis, L_i and $\hat{L}$ follow an arbitrary distribution and mutually independent, yet not necessarily identically distributed. Since PDCs are equipped with latency calculation function, it can be safely assumed that we know the stochastic information of all latencies. Furthermore, we have $0 < j \leq n$ deadline τ_j as stopping time. Then, at any moment t, the total number of received measurements is given by

$$N(t) = \sum_{i=1}^{m} \mathbf{1}_{L_i < t} \tag{1}$$

where $\mathbf{1}_{\{\}}$ is the indicator function.

When $t = \tau_j$, we define the reward function as

$$R(\tau_j) = |N(\tau_j) - N(\tau_{j-1})|F_{\hat{L}}(D - \tau_j) - c. \tag{2}$$

In the above equation, $F_{\{\}}$ is the cumulative distribution function of random variable $\hat{L}$. It can be considered as a discount factor for the received yet unsent measurements. The later it sends, the less reward we will get. c is a constant cost for sending one combined packet from PDC to SCADA, which could force the PDC to send with patience and save the bandwidth. Here we ignored the process time for PDC since it can be considered as a fixed time for specific equipment and can be easily incorporated into the requirement of D.

If we have n shooting times, then the total reward is given as

$$R = \sum_{j=1}^{n} [|N(\tau_j) - N(\tau_{j-1})|F_{\hat{L}}(D - \tau_j) - c]. \tag{3}$$

To simplify the notation, we let $N(\tau_0) = 0$ and $\tau_0 = 0$. By maximizing reward R,

$$R^* = \arg \max_{0 < \tau_1 < \tau_2 \cdots < \tau_n} R(\tau_1, \tau_2, \ldots, \tau_n) \tag{4}$$

provides the best aggregation strategy for PDC. Please notice that we neither assign nor limit the numbers of shooting times n in the PDC; however it should be automatically determined by the algorithm. The decision of these shooting times are similar to a sequential decision problem from the view of PDC over time. Therefore it is ready to be optimally solved by stochastic dynamic programming.

3.2   Stochastic Dynamic Programming

We consider a discrete time model for the PDC queue in which each time interval last t_s. The PDC will make a observation for the states. For simplicity, we let $K = \frac{D}{t_s}$ be an integer and time $t \in \{0, 1, \ldots, K\}$. The system has two states during the process. First one is the PMU measurements $N(t)$ received by time t. The other state $S(t)$ is the record of measurements that have been sent by PDC. We use $X(t)$ to represent the tuple $(N(t), S(t))$ concisely.

Action and Strategy: During each time slot, PDC need to make a decision of whether it transmits the messages received by then. We use

$$a(t) = u(X(t), t) = \begin{cases} 1 \text{ if PDC reach a stopping time.} \\ 0 \qquad\qquad \text{otherwise.} \end{cases}$$

as the action. Therefore, the number of shooting is determined by how many $a(t)$ is non-zero over one period.

Dynamic of System States: $N(t)$ could be considered a stochastic process and it can be simplified as a Markov chain with transition probability $P(N(t+1)|N(t))$ under certain assumption. $S(t) = a(t)N(t) + (1 - a(t)) \times S(t-1)$. It will not change until an shooting action is carried out and $S(\tau_j)$ is updated as $N(\tau_j)$.

Benefit Function: Given all states and action of t, we define the gain function in each time slot.

$$C(X(t), a(t)) = a(t)[(N(t) - S(t))F_{PDC}(D - t) - c]$$

Bellman's Function: The key challenge of this scheduling algorithm is how to choose the shooting time given no knowledge of the evolution of states in the future. The action made before current time will have an impact on the expectation of future gain. Given above elements, we have following expectation form of Bell function.

$$J_t(X(t)) = \max_{a(t) \in \{0,1\}} (C(X(t), a(t)) + \mathbf{E}[J_{t+1}(X(t+1))|X(t)])$$

In the next section, we will analyze the performance in the simulation.

4   Numerical Results

Given the problem formulation, we conducted the numeral simulation to verify the performance. Without loss of generality, we assume one PDC between m PMUs and SCADA. We applied Monte Carlo method to generate the random latencies to calculate the average packet loss rates and average numbers of packet that have been sent in one period. We compared our method with other two strategies — single optimal stopping aggregation method proposed in [3] and the scheme of multiple fixed shooting times. The multiple fixed shooting times $T_{fixed}(i)$ are determined by number H of the average packets that have been sent. First we need to find the maximum positive integer $\lfloor H \rfloor$ in different scenarios, and then $T_{fixed}(i) = \frac{D}{\lfloor H \rfloor + 1} \times i$ where $0 < i \leq \lfloor H \rfloor$. All the latencies L_i and $\hat{L}$ are modeled as independent random variables following exponential distribution with parameter λ. The parameters we used in the simulation are listed in Table 1. We compare these approaches by varying one of these simulation parameters. Before we gave detailed results, some general results are summarized based on all cases.

4.1   General Result

As can be seen from all scenarios, our approach outperforms other approaches. However, in most cases, the scheme of multiple fixed shooting times outperforms the one optimal stopping time, if the average sending times are larger than 2. It shows that the packets arrived later than simple deadline should not be abandoned, as long as it does not reach the physical distance limitation. All these

Table 1. Parameters of simulation setup

Cost	0.6
λ_1	5
λ_2	5
PMU Number	15
Latency D	30 ms

simulations demonstrate that there is still great room for improvement on packet loss in synchrophasor networks.

PMU Number: We varied PMU number m in the simulation scenario but the packet loss rate barely changed as showed in Fig. 3. However, from the average packet number in the figure, we learned that it increases with number of PMU data steams. Considering more PMUs could lead to more processing time in PDC node, it might be efficient to limit the number of PMUs within a small range to reduce the bandwidth cost and preserve more time for the net latency allowance.

Cost per Packet: This coefficient put a penalty on the total shooting time. By adjusting the cost per packet c in the objective function, we could see that the packet success rate decreases with the cost, which fits our intuition. At the same time, the average shooting time also decreases, which reduces the bandwidth cost. Based on the simulation result, we could guarantee a better packet loss rate by reserving a higher bandwidth.

Latency Distribution: In this part, we change the parameter λ of exponential distribution and result is showed in Figs. 5 and 6. Since the total latency allowance D is fixed, the packet loss rate is negatively correlated with λ as shown in the figures. Since the expected latency increases with λ, we expect that the total budget becomes tighter and thus, more packets can not reach the sink in time. However, the average shooting time also increases with λ in general. We believe that PDC is trying to send more packets at the latter segment of the period. However, the success rate becomes lower due to the increasing expected latency between PDC and PMUs. In a nutshell, an over-tight total latency will waste both bandwidth and PMU measurements (Fig. 4).

Total Latency Allowance D: In Fig. 7, three methods finally reach the same level as we relax on the requirement of D. However, our method still beats others under more stringent situations in which total latency allowances are very limited.

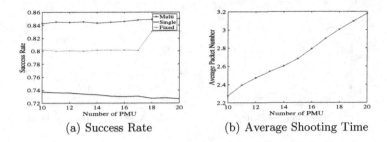

(a) Success Rate

(b) Average Shooting Time

Fig. 3. The impact of number of PMUs

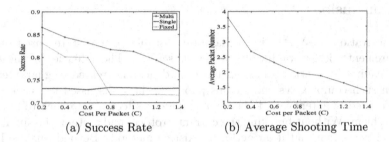

(a) Success Rate

(b) Average Shooting Time

Fig. 4. The impact of cost per packet

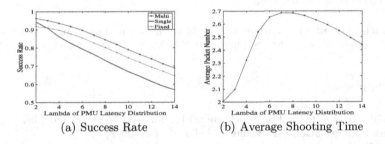

(a) Success Rate

(b) Average Shooting Time

Fig. 5. The impact of distribution λ_1 of PMU latency

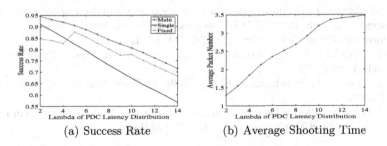

(a) Success Rate

(b) Average Shooting Time

Fig. 6. The impact of distribution λ_2 of PDC latency

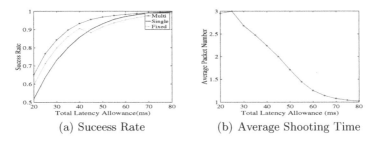

(a) Sucess Rate   (b) Average Shooting Time

Fig. 7. The impact of total latency D

5   Conclusion

The motivation of WAMS is to maintain a stable system state based on the measurements of geographically dispersed sensors. The latencies of measurements cannot be overlooked and need to be delicately treated to guarantee the QoS in such complex system. In this paper, we have discussed the packet loss issues caused by latency, and offered a simple yet effective approach to mitigate this problem. Without breaking the current protocol and system, the simulation results have shown that it outperforms existing methods. Beyond this result, we will further study more general cases with non-uniform distributions, where the number of states grows exponentially and some approximations are needed to avoid the curse of dimension.

Acknowledgment. The authors would like to thank the support of National Science Foundation under grants ECCS-1407679, CNS-1525226, CNS-1525418, CNS-1543830.

References

1. Phadke, A.G., Thorp, J.S.: Synchronized Phasor Measurements and Their Applications. Springer Science & Business Media, Boston (2008)
2. Gao, P., Wang, M., Ghiocel, S.G., Chow, J.H., Fardanesh, B., Stefopoulos, G.: Missing data recovery by exploiting low-dimensionality in power system synchrophasor measurements. IEEE Trans. Power Syst. **31**(2), 1006–1013 (2016)
3. He, M., Zhang, J.: Deadline-aware concentration of synchrophasor data: an optimal stopping approach. In: 2014 IEEE International Conference on Smart Grid Communications (SmartGridComm), pp. 296–301. IEEE (2014)
4. IEEE Standard for Synchrophasor Measurements for Power Systems. IEEE C37.118.1-2011 (2011)
5. IEEE Standard for Synchrophasor Data Traffic for Power Systems. IEEE C37.118.2-2011 (2011)
6. IEEE Guide for Phasor Data Concentrator Requirements for Power System Protection, Control, and Monitoring. IEEE C37.244 (2013)

Energy Trading Game for Microgrids Using Reinforcement Learning

Xingyu Xiao, Canhuang Dai, Yanda Li, Changhua Zhou, and Liang Xiao[✉]

Department of Communication Engineering, Xiamen University,
361005 Xiamen, China
lxiao@xmu.edu.cn

Abstract. Due to the intermittent production of renewable energy and the time-varying power demand, microgrids (MGs) can exchange energy with each other to enhance their operational performance and reduce their dependence on power plants. In this paper, we investigate the energy trading game in smart grids, in which each MG chooses its energy trading strategy with its connected MGs and power plants according to the energy generation model, the current battery level, the energy demand, and the energy trading history. The Nash equilibria of this game are provided, revealing the conditions under which the MGs can satisfy their energy demands by using local renewable energy generations. In a dynamic version of the game, a Q-learning based strategy is proposed for an MG to obtain the optimal energy trading strategy with other MGs and the energy plants without being aware of the future energy consumption model and the renewable generation of other MGs in the trading market. We apply the estimated renewable energy generation model of the MG and design a hotbooting technique to exploit the energy trading experiences in similar scenarios to initialize the quality values in the learning process to accelerate the convergence speed. The proposed hotbooting Q-learning based energy trading scheme significantly reduces the total energy that the MGs in the smart grid purchase from the power plant and improves the utility of the MG.

Keywords: Energy trading · Game theory · Reinforcement learning · Smart grids

1 Introduction

As important entities in smart grids, microgrids (MGs) are small-scale power supply networks that consist of renewable energy generators, such as wind turbines and solar panels, local electrical consumers and energy storage devices [1]. Each MG is aware of the local energy supply and the demand profiles of other

This research was supported in part by National Natural Science Foundation of China under Grant 61671396 and the CCF-Venustech Hongyan Research Initiative (2016-010).

© ICST Institute for Computer Sciences, Social Informatics and Telecommunications Engineering 2017
L. Duan et al. (Eds.): GameNets 2017, LNICST 212, pp. 131–140, 2017.
DOI: 10.1007/978-3-319-67540-4_12

MGs and the nearby power plant such as the energy selling prices using wireless networks [2]. Therefore, microgrids with extra energy can sell energy to other microgrids with insufficient energy to reduce their dependence on the energy generated by the power plants with fossil fuel and save the long-distant energy transmission loss.

Game theory is an important tool to study the energy trading in smart grids [3–8]. For example, the energy demand of consumers and the response of utility companies are formulated as a Stackelberg game in [4], yielding a reserve power management scheme to decide energy trading price. The energy trading of a power facility controller to buy energy from the power plant and multiple residential users was studied in [6], which yields a charging-discharging strategy to minimize the total energy purchase cost. The energy exchange game for MGs formulated in [7] analyzes the subjectivity decision of end-users in the energy exchange with prospect theory. The energy exchange game developed in [8] addresses energy cheating with the indirect reciprocity principle.

However, to our best knowledge, the game theoretical study on energy trading among multiple MGs with heterogeneous and autonomous operators and renewable energy supply are still open issues. In this paper, we formulate the energy exchange interactions among interconnected MGs and the power plant as an energy trading game, in which each MG chooses the amount of energy to sell to or purchase from the connected MGs and the power plants in the smart grid based on its battery level, the energy generation model and the trading history. The MGs negotiate with each other on the amount of trading energy according to the time-varying renewable energy generation and power demand of the MGs. The energy generation model such as [13] is incorporated in the energy trading game to estimate the renewable energy generation. The Nash equilibrium (NE) of this game is derived, disclosing the conditions that the MGs are motivated to provide their extra renewable energy to other MGs and purchase less energy from the power plants.

Reinforcement learning techniques, such as Q-learning can be used by smart grids to manage the energy storage and generation. For example, a temporal difference-learning based storage control scheme proposed in [9] for the residential users can minimize the electric bill without knowing the power conversion efficiencies of the DC/AC converters. The Q-learning algorithm based heterogeneous storage control system with multiple battery types proposed in [10] improves the system efficiency. In a two-layer Markov model based on reinforcement learning investigated in [11], generators choose whether to participate in the next days generation process in the power grid to improve both the day-ahead and real-time reliability. However, these works focus on the energy storage and generation rather than the energy trading among the MGs.

In this paper, a Q-learning based energy trading strategy is proposed for the MG to derive the optimal policy via trial-and-errors without being aware of the energy demand model and the storage level of other MGs in the dynamic game. To accelerate the learning speed, we exploit the renewable energy generation model in the learning process and design a hotbooting technique that applies the trading experiences in similar smart grid scenarios to initialize the quality values of the Q-learning algorithm at the beginning of the game. Simulation results show that the hotbooting Q-learning based energy trading scheme further promotes the energy trading among the connected MGs in a smart grid, reduces the reliance on the energy from the power plants, and significantly improves the utility of the MGs.

The rest of this paper is organized as follows: The energy trading game is formulated in Sect. 2, and the NE of the game is provided in Sect. 3. A hotbooting Q-learning based energy trading strategy is proposed for the dynamic game in Sect. 4. Simulation results are provided in Sect. 5, and conclusions are drawn in Sect. 6.

2 Energy Trading Game

We consider an energy trading game consisting of N MGs that are connected with each other and a power plant in the main grid via a substation. Each MG is equipped with renewable power generators, active loads, electricity storage devices, and the power transmission lines connecting with other MGs and the power plant. A microgrid has energy supply from other microgirds, the power plant, and local renewable energy generators based on wind, photovoltaic, biomass, and tidal energy.

The renewable energy generation such as wind power is local-independent, intermittent and time-varying. The amount of the energy generated by renewable power generators in MG i at time k denoted by $g_i^{(k)}$ can be estimated via the power generation history and the modeling method such as [13], yielding an estimated amount of the generated power denoted by $\hat{g}_i^{(k)}$. For simplicity, the estimation error regarding $g_i^{(k)}$ is assumed to follow a uniform distribution, given by

$$g_i^{(k)} - \hat{g}_i^{(k)} \sim G \cdot \mathrm{U}(-1, 1), \tag{1}$$

where G is the maximum estimation error.

In a smart grid, the energy trading interaction among the MGs can be formulated as an energy trading game that consists of N players. The amount of energy that MG i intends to sell to (or buy from) MG j before the bargaining is denoted by $x_{ij}^{(k)}$, which is chosen by MG i based on the observed state of the smart grid, such as its battery level, the energy trading prices, and its current energy production, and the energy demand. The trading strategy of MG i at time k is denoted by $\boldsymbol{x}_i^{(k)} = [x_{ij}^{(k)}]_{1 \leq j \leq N} \in \boldsymbol{X}$, where $\boldsymbol{X}$ is the feasible action set of the MGs and $x_{ii}^{(k)}$ is the amount of energy that MG i intends to trade with the power plant. If $x_{ij}^{(k)} > 0$, MG i intends to sell its extra energy to MG j or the power plant. If $x_{ij}^{(k)} < 0$, MG i aims to buy energy.

Note that sometimes two MGs intend to sell energy to each other at the same time, i.e., $x_{ij}^{(k)} x_{ji}^{(k)} > 0$. This problem has to be addressed with the energy trading bargaining. The resulting actual trading strategy of MG i at time k is denoted by $\boldsymbol{y}_i^{(k)} = [y_{ij}^{(k)}]_{1 \leq j \leq N}$, where $y_{ij}^{(k)}$ and $y_{ii}^{(k)}$ denote the amounts of the energy sold if positive by MG i to the power plant and MG j, respectively, or the amount of the energy purchased from them if negative, with $|y_{ij}^{(k)}| \leq C$, in which C is the maximum amount of energy exchange between two MGs. The time index k is omitted, if no confusion incurs. Therefore, the actual amount of trading energy between MG i and MG j after the bargaining is based on their intention trading interactions and given by

$$
y_{ij} = \begin{cases} -\min(-x_{ij}, x_{ji}), & \text{if } x_{ij} < 0,\ x_{ji} > 0 \\ \min(x_{ij}, -x_{ji}), & \text{if } x_{ij} > 0,\ x_{ji} < 0 \\ 0, & \text{o.w.} \end{cases} \tag{2}
$$

In this way, we can ensure that $y_{ij} + y_{ji} = 0$, $\forall i \neq j$. The amount of the energy that MG i trades with the energy plant is given by

$$
y_{ii} = \sum_{1 \leq i \neq j \leq N} x_{ij} - \sum_{1 \leq i \neq j \leq N} y_{ij}. \tag{3}
$$

Energy storage devices, such as batteries, can charge energy if the load in the MG is low and discharge if the load is high. The battery level of MG i, denoted by $b_i^{(k)}$, cannot exceed the storage capacity denoted by B, with $0 < b_i^{(k)} \leq B$. The estimated amount of the local energy demand is denoted by $d_i^{(k)}$, with $0 \leq d_i^{(k)} \leq D_i$, where D_i represents the maximum amount of local energy required by MG i. The battery level of MG i depends on the amount of trading energy, the local energy generation, and the energy demand at that time. For the smart grid with N MGs, we have

$$
b_i^{(k)} = b_i^{(k-1)} + g_i^{(k)} - d_i^{(k)} + \sum_{j=1}^{N} y_{ij}^{(k)}. \tag{4}
$$

The energy gain of MG i, denoted by $G_i(b)$, is defined as the benefit that MG i obtains from the battery level b, which is nondecreasing with b with $G(0) = 0$. As the logarithmic function is widely used in economics for modeling the preference ordering of users and for decision making [4], we assume that

$$
G_i(b) = \beta_i \ln(1 + b), \tag{5}
$$

where the positive coefficient β_i represents the ability that MG i satisfies the energy demand of the users.

To encourage the energy exchange among MGs, the local market provides a lower selling price for the trade between MGs denoted by $\rho^{-(k)}$ and a higher buying price denoted by $\rho^{+(k)}$, compared with the prices offered by the power plant which are denoted by $\rho_p^{-(k)}$ and $\rho_p^{+(k)}$, respectively, i.e., $\rho^{-(k)} > \rho_p^{-(k)}$ and $\rho^{+(k)} < \rho_p^{+(k)}$.

The utility of MG i at time k, denoted by $u_i^{(k)}$, depends on the energy gain and the trading profit, given by

$$
\begin{aligned}
u_i^{(k)}(\boldsymbol{y}) =& \beta \ln \left(1 + b_i^{(k-1)} + g_i^{(k)} - d_i^{(k)} + \sum_{j=1}^{N} y_j \right) - \sum_{j\neq i}^{N} y_j \left(\mathrm{I}(y_j \leq 0)\rho^{-(k)} \right. \\
& \left. + \mathrm{I}(y_j > 0)\rho^{+(k)} \right) - y_i \left(\mathrm{I}(y_i \leq 0)\rho_p^{-(k)} + \mathrm{I}(y_i > 0)\rho_p^{+(k)} \right),
\end{aligned}
\tag{6}
$$

where $\mathrm{I}(\cdot)$ be an indicator function that equals 1 if the argument is true and 0 otherwise.

3   NE of the Energy Trading Game

We first consider the NE of the energy trading game with $N = 2$ MGs, which is denoted by $\boldsymbol{x}_i^* = [x_{ij}^*]_{1\leq j\leq 2}$. Each MG chooses its energy trading strategy at the NE state to maximize its own utility, if the other MG applies the NE strategy. By definition, we have

$$
u_1(\boldsymbol{x}_1^*, \boldsymbol{x}_2^*) \geq u_1(\boldsymbol{x}_1, \boldsymbol{x}_2^*), \forall \boldsymbol{x}_1 \in X
\tag{7}
$$

$$
u_2(\boldsymbol{x}_1^*, \boldsymbol{x}_2) \leq u_2(\boldsymbol{x}_1^*, \boldsymbol{x}_2^*), \forall \boldsymbol{x}_2 \in X.
\tag{8}
$$

Theorem 1. *The energy trading game with $N = 2$ microgrids and a power plant has an NE $(\boldsymbol{x}_1^*, \boldsymbol{x}_2^*)$ given by*

$$
\boldsymbol{x}_1^* = \left[0, \frac{\beta}{\rho} - 1 - b_1^{(k-1)} - g_1^{(k)} + d_1^{(k)} \right]
\tag{9}
$$

$$
\boldsymbol{x}_2^* = \left[\frac{\beta}{\rho - 1} - 1 - b_2^{(k-1)} - g_2^{(k)} + d_2^{(k)}, 0 \right],
\tag{10}
$$

if

$$
\left\{
\begin{aligned}
& \rho^- = \rho^+ = \rho_p^+ - 1 = \rho_p^- + 1 = \rho \tag{11a} \\
& 0 < \frac{\beta}{\rho} - 1 - b_1^{(k-1)} - g_1^{(k)} + d_1^{(k)} \\
& \qquad < -\frac{\beta}{\rho - 1} + 1 + b_2^{(k-1)} + g_2^{(k)} - d_2^{(k)} \tag{11b} \\
& |x_{12}| \leq |x_{21}| \tag{11c} \\
& x_{12} > 0, \; x_{21} < 0. \tag{11d}
\end{aligned}
\right.
$$

Proof. If (11) holds, by (2) and (3), we have $x_{11} = x_{22} = 0$ and $y_{12} = \min(x_{12}, -x_{21}) = x_{12}$, and thus (6) can be simplified into

$$
u_1(\boldsymbol{x}_1, \boldsymbol{x}_2^*) = \beta \ln \left(1 + b_1^{(k-1)} + g_1^{(k)} - d_1^{(k)} + x_{12} \right) - x_{12}\rho,
\tag{12}
$$

$$
u_2(\boldsymbol{x}_1^*, \boldsymbol{x}_2) = \beta \ln \left(1 + b_2^{(k-1)} + g_2^{(k)} - d_2^{(k)} + x_{21} \right) - x_{21}(\rho - 1) + x_{12}^*.
\tag{13}
$$

Thus, we have

$$\frac{du_1(\boldsymbol{x}_1, \boldsymbol{x}_2^*)}{dx_{12}} = \frac{\beta}{1 + b_1^{(k-1)} + g_1^{(k)} - d_1^{(k)} + x_{12}} - \rho, \tag{14}$$

and

$$\frac{d^2 u_1(\boldsymbol{x}_1, \boldsymbol{x}_2^*)}{dx_{12}^2} = -\frac{\beta}{\left(1 + b_1^{(k-1)} + g_1^{(k)} - d_1^{(k)} + x_{12}\right)^2} < 0, \tag{15}$$

indicating that $u_1(\boldsymbol{x}_1, \boldsymbol{x}_2^*)$ is convex in terms of $\boldsymbol{x}_1$. Thus the solution of $du_1(\boldsymbol{x}_1, \boldsymbol{x}_2^*)/dx_{12} = 0$ is given by (10). Thus $u_1(\boldsymbol{x}_1, \boldsymbol{x}_2^*)$ is maximized by $\boldsymbol{x}_1^*$ in (9), indicating that (7) holds. Similarly, we can prove that (8) holds.

Corollary 1. *At the NE of the energy trading game with $N = 2$ MGs if (11) hold, MG 1 buys y_{12}^* amount of energy from MG 2, and the latter sells $-y_{22}^*$ energy to the power plant, with*

$$y_{12}^* = \frac{\beta}{\rho} - 1 - b_1^{(k-1)} - g_1^{(k)} + d_1^{(k)} \tag{16}$$

$$- y_{22}^* = \beta \frac{2\rho - 1}{\rho(\rho - 1)} + 2 + \sum_{i=1}^{N} \left(b_i^{(k-1)} + g_i^{(k)} - d_i^{(k)} \right), \tag{17}$$

and the utility of MG 1 and that of MG 2 are given respectively by

$$u_1 = \beta \left(\ln \frac{\beta}{\rho} - 1 \right) + \rho \left(1 + b_1^{(k-1)} + g_1^{(k)} - d_1^{(k)} \right) \tag{18}$$

$$u_2 = \beta \left(\ln \frac{1}{\rho - 1} - 1 + \frac{1}{\rho} \right) + \rho \left(1 + b_2^{(k-1)} + g_2^{(k)} - d_2^{(k)} \right)$$

$$- 2 - \sum_{i=1}^{2} \left(b_2^{(k-1)} + g_2^{(k)} - d_2^{(k)} \right). \tag{19}$$

4   Energy Trading Based on Hotbooting Q-Learning

The repeated interactions among N MGs in a smart grid can be formulated as a dynamic energy trading game. The amounts of the energy that MG i trades with the power plant and other MGs impact on its future battery level and the future trading decisions of other MGs as shown in (2) and (4). Thus the next state observed by the MG depends on the current energy trading decision, indicating a Markov decision process. Therefore, an MG can use Q-learning to derive the optimal trading strategy without knowing other MGs' battery levels and energy demand models in the dynamic game. More specifically, the amount of the energy that MG i intends to sell or purchase in the smart grid at time k, i.e. $\boldsymbol{x}_i^{(k)}$, is chosen based on its quality function or Q-function denoted by $Q_i(\cdot)$, which describes the expected discounted long-term reward for each state-action pair. The state observed by MG i at time slot k, denoted by $s_i^{(k)}$, consists of the current local energy demand, the estimated amount of the renewable energy generated at time k and the previous battery level of the MG, i.e., $s_i^{(k)} = \left[d_i^{(k)}, \hat{g}_i^{(k)}, b_i^{(k-1)} \right]$.

The value function $V_i(s)$ is the maximal Q function over the feasible actions at state s. The Q function and the value function of MG i are updated, respectively, by the following:

$$Q_i\left(s_i^{(k)}, x_i^{(k)}\right) \leftarrow (1-\alpha)Q_i\left(s_i^{(k)}, x_i^{(k)}\right) + \alpha\left(u_i^{(k)} + \gamma V_i\left(s_i^{(k+1)}\right)\right) \tag{20}$$

$$V_i\left(s_i^{(k)}\right) = \max_{x \in X} Q_i\left(s_i^{(k)}, x\right), \tag{21}$$

where $\alpha \in (0,1]$ is the learning rate representing the weight of current experience in the learning process, and the discount factor $\gamma \in [0,1]$ indicates the uncertainty of the microgrid regarding the future utility.

The standard Q-learning algorithm initializes the Q-function with an all-zero matrix, which is usually not the optimal value and thus degrades the learning performance at the beginning. Therefore, we design a hotbooting technique to initialize the Q-value based on the training data obtained from the large-scale experiments performed in similar smart grid scenarios. This saves the random explorations at the beginning of the game and thus accelerates the convergence rate. More specifically, we perform I similar energy trading experiments before the start of the game, as shown in Algorithm 1.

Algorithm 1. Hotbooting process for MG i.

Initialize α, γ, $Q_i^*(s_i, x_i) = 0$, and $V_i^*(s_i) = 0$, $\forall s_i, x_i$
Set $b_i^{(0)} = 0$
For $t = 1, 2, \cdots, I$
 Emulate a similar energy trading scenario for N MGs
 For $k = 1, 2, \cdots, K$
  Observe $\hat{g}_i^{(k)}$ and $d_i^{(k)}$
  Obtain state $s_i^{(k)} = \left[d_i^{(k)}, \hat{g}_i^{(k)}, b_i^{(k-1)} \right]$
  Choose $x_i^{(k)} \in X$ via Eq. (22)
  For $j = 1, 2, \cdots, N$
   If $j \neq i$
    Negotiate with MG j to obtain $y_{ij}^{(k)}$ via (2)
    Sell or purchase $|y_{ij}^{(k)}|$ amount of the energy to or from MG j
   Else
    Calculate $y_{ii}^{(k)}$ via (3)
    Sell or purchase $|y_{ii}^{(k)}|$ amount of the energy to or from the power plant
   End if
  End for
  Obtain $u_i^{(k)}$
  Observe $b_i^{(k)}$
  Calculate $Q_i^*\left(s_i^{(k)}, x_i^{(k)}\right)$ via (20)
  Calculate $V_i^*\left(s_i^{(k)}\right)$ via (21)
 End for
End for

Algorithm 2. Hotbooting Q-learning based energy trading of MG i.

Initialize α, γ, $Q_i=Q_i^*$, and $V_i=V_i^*$
Set $b_i^{(0)} = 0$
For $k = 1, 2, \cdots$
    Estimate $\hat{g}_i^{(k)}$ and $d_i^{(k)}$
    Obtain state $\boldsymbol{s}_i^{(k)} = \left[d_i^{(k)}, \hat{g}_i^{(k)}, b_i^{(k-1)}\right]$
    Select the trading strategy $\boldsymbol{x}_i^{(k)}$ via Eq. (22)
    For $k = 1, 2, \cdots, K$
      If $j \neq i$
        Negotiate with MG j to obtain $y_{ij}^{(k)}$ via (2)
        Sell or purchase $|y_{ij}^{(k)}|$ amount of the energy to or from MG j
      Else
        Calculate $y_{ii}^{(k)}$ via (3)
        Sell or purchase $|y_{ii}^{(k)}|$ amount of the energy to or from the power plant
      End if
    End for
    Obtain $u_i^{(k)}$
    Observe $b_i^{(k)}$
    Update $Q_i\left(\boldsymbol{s}_i^{(k)}, \boldsymbol{x}_i^{(k)}\right)$ via Eq. (20)
    Update $V_i\left(\boldsymbol{s}_i^{(k)}\right)$ via Eq. (21)
End for

To balance the exploitation and exploration in the learning process, an ϵ-greedy policy is applied to choose the amount of the energy to trade with other MGs and the energy plant, i.e., $\boldsymbol{x}_i^{(k)}$ is given by

$$\Pr(\boldsymbol{x}_i^{(k)} = \boldsymbol{\Theta}) = \begin{cases} 1 - \epsilon, & \boldsymbol{\Theta} = \arg\max_{\boldsymbol{x}\in\boldsymbol{X}} Q_i\left(\boldsymbol{s}_i^{(k)}, \boldsymbol{x}\right) \\ \frac{\epsilon}{|\boldsymbol{X}|}, & \text{o.w.} \end{cases} \tag{22}$$

MG i chooses $\boldsymbol{x}_i^{(k)}$ according to ϵ-greedy strategy and negotiates with other MGs to determine the actual amounts of the energy in the trading $\boldsymbol{y}_i^k$ according to (2). As shown in Algorithm 2, the MG observes the reward and the next state. According to the resulting utility $u_i^{(k)}$, the MG updates its Q function via (20) and (21).

5   Simulation Results

Simulations have been performed to evaluate the performance of the hotbooting Q-learning based energy trading strategy in the dynamic game with $N = 2$ MGs. In the simulation, if not specified otherwise, the energy storage capacity of each MG is $B = 4$, and the energy gain is $\beta = 8$. The local energy demands, the energy trading prices, and the renewable energy generation models of each MG in the simulations are retrieved from the energy data of microgrids in Hong kong in [13]. As benchmarks, we consider the Q-learning based trading scheme and the greedy scheme, in which each MG chooses the amount of selling/buying energy according to its current battery level to maximize its estimated immediate utility.

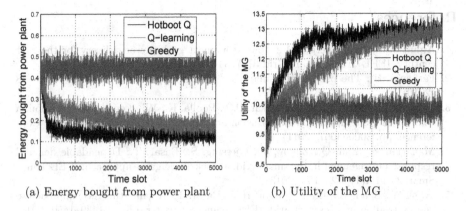

Fig. 1. Performance of the energy trading strategies in the dynamic game with $N = 2$, $B = 4$ and $\beta = 8$

As shown in Fig. 1, the proposed Q-learning based energy trading strategy outperforms the greedy strategy with less energy bought from the power plant and a higher utility of the MG. For example, the Q-learning based strategy decreases the average amount of the energy purchased from the power plant by 47.7% and increases the utility of the MG by 11.6% compared with the greedy strategy at the 1500-th time slot in the game. The performance of the Q-learning based strategy is further improved with the hotbooting technique that exploits similar energy trading experiences to accelerate the learning speed. As shown in Fig. 1, the hotbooting Q-learning based energy trading strategy decreases the amount of the energy purchased from the power plant by 33.7% and increases the utility of the MG by 9.5% compared with the Q-learning based strategy at the 1500-th time slot.

6 Conclusion

In this paper, we have formulated an MG energy trading game for smart grids and derived the NE of the game, disclosing the conditions under which the MGs in a smart grid trade with each other and reduce the dependence on the power plant. A Q-learning based energy trading strategy has been proposed for each MG to choose the amounts of the energy to trade with other MGs and the power plant in the dynamic game with time-varying renewable energy generations and power demands. The learning speed is further improved by the hotbooting Q-learning technique. Simulation results show that the proposed hotbooting Q-learning based energy trading technique improves the utility of MG and reduces the amount of the energy purchased from the power plant, compared with the benchmark strategy.

References

1. Amin, S.M., Wollenberg, B.F.: Toward a smart grid: power delivery for the 21st century. IEEE Trans. Smart Grid **3**(5), 34–41 (2005)
2. Farhangi, H.: The path of the smart grid. IEEE Power Energ. Mag. **8**(1), 18–28 (2010)
3. Baeyens, E., Bitar, E., Khargonekar, P.P., Poolla, K.: Wind energy aggregation: a coalitional game approach. In: Decision and Control and European Control Conference, pp. 3000–3007 (2011)
4. Maharjan, S., Zhu, Q., Zhang, Y., Gjessing, S., Basar, T.: Dependable demand response management in the smart grid: a stackelberg game approach. IEEE Trans. Smart Grid **4**(1), 120–132 (2013)
5. Wang, Y., Saad, W., Han, Z., Poor, H.V., Basar, T.: A game-theoretic approach to energy trading in the smart grid. IEEE Trans. Smart Grid **5**(3), 1439–1450 (2014)
6. Tushar, W., Chai, B., Yuen, C., et al.: Three-party energy management with distributed energy resources in smart grid. IEEE Trans. Ind. Electron. **62**(4), 2487–2498 (2015)
7. Xiao, L., Mandayam, N.B., Poor, H.V.: Prospect theoretic analysis of energy exchange among microgrids. IEEE Trans. Smart Grid **6**(1), 63–72 (2015)
8. Xiao, L., Chen, Y., Liu, K.R.: Anti-cheating Prosumer Energy Exchange based on Indirect Reciprocity. In: IEEE International Conference on Communication, pp. 599–604. Sydney (2014)
9. Guan, C., Wang, Y., Lin, X., Nazarian, S., Pedram, M.: Reinforcement learning-based control of residential energy storage systems for electric bill minimization. In: IEEE Consumer Communication and Networking Conference, pp. 637–642. Las Vegas (2015)
10. Qiu, X., Nguyen, T.A., Crow, M.L.: Heterogeneous energy storage optimization for microgrids. IEEE Trans. Smart Grid **7**(4), 1453–1461 (2016)
11. Dalal, G., Gilboa, E., Mannor, S.: Hierarchical decision making in electricity grid management. In: International Conference on Machine Learning, New York, pp. 2197–2206 (2016)
12. Kaelbling, L.P., Littman, M.L., Moore, A.W.: Reinforcement learning: a survey. J. Artif. Intell. Res. **4**, 237–285 (1966)
13. Wang, H., Huang, J.: Incentivizing Energy Trading for Interconnected Microgrids. IEEE Trans. Smart Grid (2016)

Comparing Customer Taste Distributions in Vertically Differentiated Mobile Service Markets

Olga Galinina[1], Leonardo Militano[2], Antonino Orsino[1(✉)], Sergey Andreev[1], Giuseppe Araniti[2], Antonio Iera[2], and Yevgeni Koucheryavy[1]

[1] Tampere University of Technology, Tampere, Finland
antonino.orsino@tut.fi

[2] University Mediterranea of Reggio Calabria, Reggio Calabria, Italy

Abstract. In this paper, we study a vertically differentiated duopoly market, where competitors (mobile service providers) offer mobile subscriptions to customers, who diversify in their preferences regarding price and quality. We consider a two-stage game where the players first select the quality and then begin a competitive process for the price or quantity, which is widely known as Bertrand or Cournot game, respectively. To capture the service provider strategy, we first introduce variable costs to improve the quality, which are linear in quality per a subscription, and then derive the market-related metrics of interest for the tractable uniform distribution of the customer's taste parameter. Further relaxing this strong assumption, we provide with a numerical procedure that helps characterize an arbitrary taste distribution as well as an arbitrary cost function. Finally, selected numerical examples report on the comparison between the uniform and the truncated exponential distribution, thus accentuating the importance of choosing an appropriate customer taste model.

1 Introduction

The telecommunications industry has already entered a new phase of its evolution, where the focus has shifted from the conventional multimedia transmission to the ubiquitous connectivity and massive traffic volumes driven by growing human demand for data as well as supported by the emerging innovations, such as the Internet of Things, wearables, and more far-fetched autonomous vehicles [1]. On this market that crossed the 100% penetration mark, competition of mobile service providers for increased market share and retention of customers becomes a vital part of their strategy.

One of the key marketing strategies for competitors to seek profitable niches is product differentiation and pricing [2]. In particular, *horizontal* differentiation refers to immeasurable distinctions in virtually identical products, such as in

This work is supported by the Finnish Cultural Foundation (Suomen Kulttuurirahasto) and by the project TT5G: Transmission Technologies for 5G.

© ICST Institute for Computer Sciences, Social Informatics and Telecommunications Engineering 2017
L. Duan et al. (Eds.): GameNets 2017, LNICST 212, pp. 141–153, 2017.
DOI: 10.1007/978-3-319-67540-4_13

design or color, which are not sufficient for the mobile service provider (SP) to attract new customers, who are willing to acquire a better level of service. In contrast to that, *vertical* market differentiation is objectively measurable and based on diverse quality levels of the products [3]. Here, customers are sensitive with respect to the relation between the quality and the price levels, and may have diverse preferences regarding it [4].

Generally, the market and pricing models have already attracted significant attention of the wireless community across a wide range of various problems, from market entrance decisions for mobile SPs [5] and competition over spectrum [6] to specific studies of social welfare in case when SPs exploit unlicensed spectrum [7]. However, to the best of our knowledge no prior work on vertical differentiation of mobile service markets has been contributed so far. In this paper, we study a duopoly model where mobile SPs first determine the specification of their offered services and then decide on the prices or the quantities of services they offer according to the Bertrand or Cournot competition models [8] (the initial market entry [9] is assumed to have been completed).

We consider both the price and the quantity competition as they lead to dissimilar equilibrium points, while there is still no consensus in past literature as to which type of competition should be preferred. We thus analyze both game models in order to reveal the dependence of the corresponding results on the optimal choice of the SP strategies, namely, whether SPs eventually offer a homogeneous product (as shown by the Cournot game) or two differentiated products (as illustrated by the Bertrand game). Since both situations may occur in the real market, one model cannot be preferred over another upfront.

Further, in modeling the mobile service markets an important role belongs to characterizing the costs of offering improved service quality. The majority of existing studies as in [9–11] assume zero or fixed quality improvement cost, as well as adopt diminishing [12] or quadratic [13] formulations. This work assumes linear costs of quality improvement per unit of product as this can be tackled easily while being close to what the SPs may experience in practice.

As an indicator of customer preferences, we adopt the standard utility function [14], where the willingness of a customer to pay for a better quality service is represented by a random *taste parameter* [14]. While most of the game-theoretical references study the formulations by example of analytically tractable but arguably unrealistic uniform distribution of the taste parameter from "poor" to "rich", we in this work offer guidance on how to handle an *arbitrary taste distribution* and an *arbitrary cost function*.

The remainder of this paper is organized as follows. In Sect. 2, we outline our system model as well as describe the two-stage game played to divide the market and set the optimal price or quantity (in Bertrand or Cournot game, respectively). **Our contributions** appear in Sects. 3 and 4, where, correspondingly, we provide analytical calculations for the conventional tractable example under the linear cost assumption and then detail our flexible numerical procedure to cope with an arbitrary formulation. Finally, we provide numerical comparison of the two considered options based on representative examples.

2   System Model

In this work, we study a *vertically differentiated* mobile service market under the simplifying assumption of two operating SPs (*service providers*). In our formulation, the SP i may be characterized by a pair "price-quality" (p_i, s_i) and offer an unconstrained number of mobile subscriptions, each of which guaranteeing the announced mobile service quality s_i for the price p_i. Thus offered subscriptions (e.g., SIM-cards) may be purchased by a potentially large number of consumers, hereinafter named *customers*. Based on their preferences, customers may select only one SP or else refrain from buying anything.

We emphasize here that the products on a vertically differentiated market (in our case, subscriptions) may differ in both their quality and price. Moreover, the customers are *not identical* in their preferences due to diverse taste or budget restrictions, which results in varying willingness to pay for the offer [4, 15].

2.1   Characterization of the Customers

Utility Function of Customers. For differentiated markets, it is typically assumed that all of the customers agree on ranking the mobile service offers (subscriptions) in the order of quality preference according to some utility function based on a *taste parameter* [13]. The taste parameter θ reflects the customer's preference i.e., the more a customer agrees to pay for a better quality service – the higher the parameter θ becomes. In our study, we adopt the following utility function of θ [15], given the price p_i and the quality s_i offered by the SP i:

$$U(\theta, s_i, p_i) = \theta \cdot s_i - p_i, \tag{1}$$

where, $s_i = s(T_i)$ is an increasing quality function of data volume or rate T_i guaranteed by the SP. The function $s(T_i)$ is typically non-linear and often represented in literature by a logarithmic dependence, but may also be replaced by another, more appropriate choice.

Strategy of Customers. All of the customers are assumed to be *rational* i.e., the strategy of any customer is to maximize its utility $U(\theta, s, p)$ by choosing *exactly one* subscription of the SP i characterized by a pair (p_i, s_i) or, alternatively, refraining from buying anything at all. We note that zero utility value $U(\theta, s_i, p_i) \leq 0$ is equivalent to not purchasing the product i, and the case $U(\theta, s_i, p_i) = U(\theta, s_j, p_j)$ yields customer's indifference to buying product i or j.

Distribution of Customers. In order to be able to apply the Cournot competition model, we further assume that the considered market is not covered i.e., there always are customers who never participate [11, 15]. Therefore, θ should be distributed over the interval $[0, \theta_{max}]$, where θ_{max} corresponds to customers able to pay the most for a better quality. We assume that within this interval θ is distributed according to a certain probability density $h_\theta(\theta)$. Below, we compare two distinct distributions $h_\theta(\theta)$: the conventional and analytically tractable uniform distribution as well as the more realistic truncated exponential distribution, for which a numerical solution may be produced.

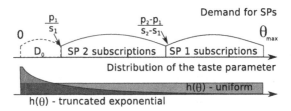

Fig. 1. Illustration of the target market structure.

2.2 Characterization of the SPs

Demand of the SPs. Without loss of generality, we reorder our SPs such that $s_1 \geq s_2$. Due to the assumption on the rationality of customers, prices should also be rearranged in the non-decreasing order $p_1 \geq p_2$. For the fixed price and quality levels, we may obtain the following points of indifference for a tagged customer [13]:

- point of indifference to buying or not buying the service of the SP 2 is denoted by the parameter $\theta_{\varnothing,2} = \frac{p_2}{s_2}$ (follows from $U(\theta, s_2, p_2) = 0$),
- point of indifference to buying the service of the SP 2 or of the SP 1 corresponds to the parameter $\theta_{2,1} = \frac{p_1 - p_2}{s_1 - s_2}$ (follows from $U(\theta, s_1, p_1) = U(\theta, s_2, p_2)$).

The demand of the SPs may then be established as $D_1(\mathbf{s}; \mathbf{p}) = \int_{\theta_{2,1}}^{\theta_{\max}} h(\theta) d\theta$ and $D_2(\mathbf{s}; \mathbf{p}) = \int_{\theta_{\varnothing,2}}^{\theta_{2,1}} h(\theta) d\theta$, where $h(\theta)$ is the probability density function (PDF) of the taste parameter θ.

Profit of the SPs. When making their decisions, the SPs abide by the principle of maximizing their *profit*, which is determined by the financial flow from the subscribed customers and depends on the structure of the costs. We assume that *linear costs* are incurred when improving the claimed quality s_i *per user*, so that the SP would be ready to support the respective quality of service (QoS) level for its subscribed customers. Hence, the total costs depend both on the number of served customers and on the selected quality level. These costs may reflect, for example, the initial investments into a fixed-term spectrum lease and/or the amounts of spectrum that could be resold (as e.g., in [16] or [17]).

Further, our profit function may be written as $\Pi_i(\mathbf{s}, \mathbf{p}) = D_i(\mathbf{s}, \mathbf{p})(p_i - \nu s_i)$, where ν is the quality cost coefficient. The latter may be roughly estimated from the value of the spectrum license costs to support the announced QoS, normalized by unit time as well as the total number of customers in the region of interest. We note that our assumption on the linear costs is relaxed in Sect. 4 and replaced by another suitable function.

2.3   Two-Stage Differentiated Market Game

In this work, we model *both alternatives*: the price and the quantity competition, which are known as the Bertrand and Cournot competition models, correspondingly. We focus on a differentiated market game with the following *two phases*:

1. In the first phase, both SPs select the quality level s_i (equivalent to e.g., a data rate package with the announced throughput). Importantly, at this stage the SPs are aware of each other, but make their decisions sequentially.
2. Second, given the fixed quality level s_i the SPs compete in price or, alternatively, in quantity. More specifically, in the Bertrand game the SPs decide on the prices $p_i, i = 1, 2$ that are announced to the customers purchasing their subscriptions. In contrast to that, in the Cournot game the SPs decide on the quantity, which in our modeling translates into the number of subscribed customers or, equivalently, sold subscriptions.

Further, we aim at determining the Nash equilibrium of our game and to do so we apply the principle of *backward induction*. Accordingly, we begin by finding an equilibrium for the second phase (price/quantity competition for the fixed levels of s_i) and then obtain the optimal values of s_i which are selected in the first phase.

3   Conventional Example: Uniform Taste Distribution

In this section, we consider a tractable example of the customer taste distribution $h(\theta)$, namely, a uniform distribution $h_U(\theta) = \frac{1}{\theta_{\max}}$ over the said interval $[0, \theta_{\max}]$ and thus the expressions for the demand may be rewritten as:

$$D_1(\mathbf{s}; \mathbf{p}) = \frac{1}{\theta_{\max}}(\theta_{\max} - \theta_{2,1}), \quad D_2(\mathbf{s}; \mathbf{p}) = \frac{1}{\theta_{\max}}(\theta_{2,1} - \theta_{\varnothing,2}). \tag{2}$$

In what follows, we consider the Bertrand price competition and the Cournot quantity competition models separately for both options.

3.1   Bertrand Price Competition for the Uniform Distribution

In the Bertrand game, the SP selects its own price p_i in order to maximize the profit $\Pi_i(\mathbf{s}; \mathbf{p}) = D_i(\mathbf{s}; \mathbf{p})(p_i - \nu s_i)$ for the selected quality function values s_i. By differentiating Π_i over p_i, one may calculate the optimal prices (can be verified for $\nu = 0$ by [13]) for the fixed levels of quality, while the solution is readily obtained as follows:

$$p_1^*(\mathbf{s}) = s_1 \frac{2\theta_{\max}(s_1 - s_2) + v(2s_1 + s_2)}{4s_1 - s_2}, \quad p_2^*(\mathbf{s}) = s_2 \frac{\theta_{\max}(s_1 - s_2) + 3s_1 v}{4s_1 - s_2}. \tag{3}$$

It can be easily demonstrated that the latter is a unique point of maximum for $0 \leq s_2 \leq s_1$, which is achieved during the price competition if all of the participants maximize their profits. The solution (3) represents a result of long-term price adjustment.

At the next step of our backward induction, we consider the first phase of the game, when the SPs select the quality s_i. Each of them maximizes the function $\Pi_i(\mathbf{s})$ over the only varying argument s_i, where:

$$\Pi_1(\mathbf{s}) = 4s_1^2 \frac{(\theta_{\max}-\nu)^2(s_1-s_2)}{\theta_{\max}(4s_1-s_2)^2}, \quad \Pi_2(\mathbf{s}) = s_1 s_2 \frac{(\theta_{\max}-\nu)^2(s_1-s_2)}{\theta_{\max}(4s_1-s_2)^2}. \tag{4}$$

The first-order condition of maximum for these two independent optimization problems may be formulated as follows:

$$\frac{4s_1(\theta_{\max}-\nu)^2\left(4s_1^2-3s_1s_2+2s_2^2\right)}{\theta_{\max}(4s_1-s_2)^3} = 0, \quad \frac{s_1^2(4s_1-7s_2)(\theta_{\max}-\nu)^2}{\theta_{\max}(4s_1-s_2)^3} = 0. \tag{5}$$

Denoting $\frac{s_1}{s_2}$ as x, we may then locate the maximum points for both SPs. We note that due to the absence of roots for the first equation and the fact that $\frac{\partial \Pi_1}{\partial s_1} > 0$, the maximum is located at the border $s_1^* = s_{\max}$, while the optimal quality $s_2^* = s_{\max}\xi$, where $\xi = 4/7$ (the second-order condition of maximum $\frac{\partial^2 \Pi_2}{\partial s_2^2}\Big|_{s_1^*,s_2^*} < 0$ could be verified easily). The latter corresponds to the rule of 4/7 [11].

Theorem 1. *The obtained solution for the Bertrand game is unique and represents the Nash equilibrium.*

Proof. The proof is fairly straightforward and is based on demonstrating that the following holds:

$$\Pi_i(s_1^*, s_2^*) \geq \Pi_i(s_1, s_2^*), \quad \text{for any } s_1 < s_1^*, \\ \text{and } \Pi_i(s_1^*, s_2^*) \geq \Pi_i(s_1^*, s_2), \quad \text{for any } s_2 \neq s_2^*, \tag{6}$$

which is based on the fact that the sought points are the points of maximum for the respective functions. Uniqueness of the sought point follows from uniqueness of $\mathbf{p}^*(\mathbf{s})$ and the solution (s_1^*, s_2^*).

Substituting the sought point $(s_{\max}, \xi s_{\max})$ into the price, demand, and profit functions, we obtain the key indicators at the equilibrium point. Then, we additionally calculate the *consumer surplus* by characterizing the integral benefit of all customers as a difference between the maximum price that they could have paid for the quality s_i (i.e., θs_i) and what they actually spend (p_i):

$$CS = \int_{\theta_{1,2}}^{\theta_{\max}} (\theta s_1 - p_1)\frac{1}{\theta_{\max}}d\theta + \int_{\theta_{\varnothing,2}}^{\theta_{1,2}} (\theta s_2 - p_2)\frac{1}{\theta_{\max}}d\theta = \frac{7s_{\max}(\theta_{\max}-\nu)^2}{24\theta_{\max}}. \tag{7}$$

3.2   Cournot Quantity Competition for the Uniform Distribution

While in the Bertrand game the price p_i is controlled by the SP i and the share of connected customers is then determined through the demand function, in the Cournot game the SPs control the quantity (i.e., the number of subscriptions) and then the prices are derived through the inverted system of demand functions:

$$p_1(\mathbf{s}; \mathbf{D}) = -\theta_{\max}(D_1 s_1 - s_1 + D_2 s_2), \\ p_2(\mathbf{s}; \mathbf{D}) = -\theta_{\max}s_2(D_1 + D_2 - 1). \tag{8}$$

Substituting the above into the expression for the SP profit $\Pi_i = D_i(p_i - vs_i)$, we may establish the quantity response functions that maximize the profit for the fixed qualities s_1 and s_2:

$$D_1(\mathbf{s}) = \frac{(2s_1-s_2)(\theta_{\max}-\nu)}{\theta_{\max}(4s_1-s_2)}, \quad D_2(\mathbf{s}) = \frac{s_1(\theta_{\max}-\nu)}{\theta_{\max}(4s_1-s_2)}.$$

After substituting these functions into (8), we obtain the prices set by the SPs:

$$p_1 = s_1 \frac{2\theta_{\max}s_1 - \theta_{\max}s_2 + 2s_1\nu}{(4s_1-s_2)}, \quad p_2 = s_2 \frac{\theta_{\max}s_1 + 3s_1v - s_2\nu}{(4s_1-s_2)},$$

and, correspondingly, characterize the resulting profit:

$$\Pi_1(\mathbf{s}) = \frac{s_1(2s_1-s_2)^2(\theta_{\max}-\nu)^2}{\theta_{\max}(4s_1-s_2)^2}, \quad \Pi_2(\mathbf{s}) = \frac{s_1^2 s_2(\theta_{\max}-\nu)^2}{\theta_{\max}(4s_1-s_2)^2}. \tag{9}$$

In the second phase of the backward induction, we derive the optimal level of qualities that maximize the profit (9) by finding the stationary points of the following equations:

$$\frac{\partial \Pi_1(\mathbf{s})}{\partial s_1} = \frac{(\theta_{\max}-\nu)^2(2s_1-s_2)(8s_1^2-2s_1s_2+s_2^2)}{\theta_{\max}(4s_1-s_2)^3}, \quad \frac{\partial \Pi_2(\mathbf{s})}{\partial s_2} = \frac{(\theta_{\max}-\nu)^2 s_1^2(4s_1+s_2)}{\theta_{\max}(4s_1-s_2)^3}. \tag{10}$$

Denoting $\frac{s_1}{s_2}$ as x, we may conclude that there exists no solution $x > 1$ for (10). Since both $\frac{\partial \Pi_1(\mathbf{s})}{\partial s_1}$ and $\frac{\partial \Pi_2(\mathbf{s})}{\partial s_2} > 0$, the point of maximum is located at the right border of the interval for s, that is, $s_1^* = s_{\max}$ and $s_2^* = s_{\max}$. Therefore, we have established a candidate solution for the Cournot game and can formulate a theorem similar to the one before.

Theorem 2. *The obtained solution for the Cournot game is unique and represents the Nash equilibrium.*

Proof. The proof is easy to produce similarly to that of the above Theorem for the Bertrand game.

Since the Cournot prices and qualities are equivalent, two SPs divide the corresponding market in equal proportions, if we assume that there is no weighted preference towards a certain brand. Hence, the consumer surplus in this case may be derived as:

$$CS = \int_{\theta_{1,2}}^{\theta_{\max}} (\theta s_1 - p_1)h(\theta)d\theta = \frac{2s_{\max}(\theta_{\max}-\nu)^2}{9\theta_{\max}}. \tag{11}$$

4   Arbitrary Taste Distribution and Cost Function

In this section, we contribute an algorithm that allows for establishing an equilibrium point for an arbitrary taste distribution and cost function. As a particular example, we refer to the truncated exponential distribution:

$$h_U(\theta) = \frac{\lambda e^{-\lambda\theta}}{1-e^{-\lambda\theta_{\max}}}, \theta \in [0, \theta_{\max}], \quad H_U(\theta) = \frac{1-e^{-\lambda\theta}}{1-e^{-\lambda\theta_{\max}}}, \theta \in [0, \theta_{\max}]. \tag{12}$$

The use of the exponential shape follows from [18], where the authors analyze a real mobile service market by polling the consumers and processing the results. Further, we truncate the exponential distribution by $\theta_{\max}$ to provide a better correspondence with the parameter of the uniform distribution. Hence, the corresponding expressions for the demand may be rewritten as:

$$D_1(\mathbf{s}; \mathbf{p}) = C_0 \left(e^{-\lambda \frac{p_1 - p_2}{s_1 - s_2}} - e^{-\theta_{\max}\lambda} \right), \quad D_2(\mathbf{s}; \mathbf{p}) = C_0 \left(e^{-\frac{\lambda p_2}{s_2}} - e^{-\lambda \frac{p_1 - p_2}{s_1 - s_2}} \right),$$

where $C_0 = \frac{1}{1 - e^{-\lambda \theta_{\max}}}$ is a constant introduced for brevity. We build our numerical comparison later on in Sect. 5 on the example of the truncated exponential distribution, which we believe to better represent the properties of the target market. However, below we formulate the essential steps of our proposed procedure in a general form as well as introduce an arbitrary cost function.

4.1   Bertrand Price Competition for an Arbitrary Distribution

The profit function in its general form is defined as $\Pi_i = D_i p_i - D_i f_c(s_i)$, where $f_c(s_i)$ is the cost per a subscription represented by the twice differentiable function of quality s_i. In this general case, we therefore have:

$$\begin{aligned}
\Pi_1(\mathbf{s}; \mathbf{p}) &= \left(1 - H\left(\frac{p_1 - p_2}{s_1 - s_2} \right) \right) (p_1 - f_c(s_1)), \\
\Pi_2(\mathbf{s}; \mathbf{p}) &= \left(H\left(\frac{p_1 - p_2}{s_1 - s_2} \right) - H\left(\frac{p_2}{s_2} \right) \right) (p_2 - f_c(s_2)),
\end{aligned} \tag{13}$$

where $H(x)$ is the cumulative distribution function (CDF) of the taste parameter and $H(\theta_{\max}) = 1$. After differentiating both expressions separately by the corresponding quality variable, we obtain a condition for further optimization:

$$\begin{aligned}
\frac{\partial \Pi_1(\mathbf{s}; \mathbf{p})}{\partial p_1} &= 1 - H\left(\frac{p_1 - p_2}{s_1 - s_2} \right) - h\left(\frac{p_1 - p_2}{s_1 - s_2} \right) \frac{p_1 - f_c(s_1)}{s_1 - s_2}, \\
\frac{\partial \Pi_2(\mathbf{s}; \mathbf{p})}{\partial p_2} &= H\left(\frac{p_1 - p_2}{s_1 - s_2} \right) - H\left(\frac{p_2}{s_2} \right) - h\left(\frac{p_1 - p_2}{s_1 - s_2} \right) \frac{p_2 - f_c(s_2)}{s_1 - s_2} - h\left(\frac{p_2}{s_2} \right) \frac{p_2 - f_c(s_2)}{s_2}.
\end{aligned} \tag{14}$$

We note that an analytical solution of the system $\left(\frac{\partial \Pi_i(\mathbf{s}; \mathbf{p})}{\partial p_i} = 0 \right)_{i=1,2}$ may not always be produced for complex distribution shapes of $f_c(s_i)$. In order to follow the steps described previously in Sect. 3, for an arbitrary distribution we may apply a numerical procedure to solve the system of non-linear equations (14) for any fixed point $\mathbf{s}$, $0 < s_2 < s_1$. If the second-order condition of the local maximum holds, the obtained solution $\mathbf{p}^*(\mathbf{s})$ is set as an output of the function $FindOptimalPrices(s_1, s_2)$, which corresponds to the second phase of our game (see Algorithm 1 below).

Continuing the search of the optimal solution, we consider again the second phase (the quality competition) and maximize the profit $\Pi_i(s_1, s_2)$ by varying s_i. Importantly, the functions $\Pi_i(s_1, s_2)$ are numerical and produced by the proposed function $FindOptimalPrices(s_1, s_2)$. The optimization can be conducted via explicit search, but the following theorem simplifies the needed calculations:

Theorem 3. *Maximum of the profit function $\Pi_1(s_1, s_2)$ by $s_1 \in (0, s_{\max}]$ for the SP that makes its decision the first is always located at the point $s_{\max}$, which means that for any new SP the maximum quality yields the highest profit.*

Proof. The proof is omitted here due to the space limitations.

Employing this result, it only remains to maximize the profit of another SP $\Pi_2(s_1, s_2)$ by $s_2 \in (0, s_1]$, which is a simple one-dimensional optimization that always has a solution. The entire procedure is briefly summarized in Algorithm 1. The sought variables $(s_1^*, s_2^*; p_1^*, p_2^*)$ correspond to the *Nash equilibrium*, where no player could change its strategy (that is, price and quality for the SPs and SP choice for the customers) without decreasing its profit. Based on the obtained equilibrium, we may easily estimate the corresponding market shares D_i^*, the equilibrium profit Π_i^*, and the consumer surplus CS as provided in Sect. 5.

4.2   Cournot Quantity Competition for an Arbitrary Distribution

In order to characterize the Cournot quantity competition for an arbitrary taste distribution and cost function, we follow the steps similar to those in Sect. 3. In particular, we write down the expression for the SPs demands:

$$D_1(\mathbf{s}; \mathbf{p}) = 1 - H\left(\frac{p_1 - p_2}{s_1 - s_2}\right), \quad D_2(\mathbf{s}; \mathbf{p}) = H\left(\frac{p_1 - p_2}{s_1 - s_2}\right) - H\left(\frac{p_2}{s_2}\right), \quad (15)$$

where $H(x)$ is the CDF of the taste parameter. From the first equation, we may establish $p_1(\mathbf{D}) = F(1 - D_1)(s_1 - s_2) + p_2$, where $F = H^{-1}$ is the function inverse to the CDF. Substituting it into the second equation and calculating p_2, we may obtain the following:

$$p_1(\mathbf{D}) = F(1 - D_1)(s_1 - s_2) + p_2, \quad p_2(\mathbf{D}) = F(1 - D_2 - D_1)s_2. \quad (16)$$

We substitute this produced expression for price into the profit function $\Pi_i(\mathbf{s}; \mathbf{D}) = D_i(p_i(\mathbf{D}) - f_c(s_i))$. By analogy with Subsect. 4.1, we find the optimal prices after differentiating the profit by the demand D_i and then solving the system $\left(\frac{\partial \Pi_i(\mathbf{s}; \mathbf{D})}{\partial D_i} = 0\right)_{i=1,2}$ as:

$$\frac{\partial \Pi_1(\mathbf{s}; \mathbf{D})}{\partial D_1} = p_1(D_1) - f_c(s_1) - \frac{D_1(s_1 - s_2)}{h(1 - D_1)}, \quad \frac{\partial \Pi_1(\mathbf{s}; \mathbf{D})}{\partial D_1} = p_2(D_2) - f_c(s_2) - \frac{D_2 s_2}{h(1 - D_2 - D_1)}, \quad (17)$$

where $h(\theta)$ is the given PDF. We note that the system (17) is equivalent to (14) for the Bertrand competition. Assuming that the function *FindOptimalQuantities(s_1, s_2)* returns the solution of (17) and then replacing prices with qualities $\mathbf{D}$ in Algorithm 1, we arrive at the final equilibrium $(s_1^*, s_2^*; D_1^*, D_2^*)$ and may calculate all of the respective metrics.

Even though existence and uniqueness of the Nash equilibrium constitute an open question for different classes of distributions, in case of our truncated exponential example we can formulate the following Theorem.

Theorem 4. *For the truncated exponential distribution, there exists a unique Nash equilibrium for both the Bertrand and the Cournot game, so that the Bertrand competition results in product differentiation, while equilibrium quality for the Cournot competition is given by* $(s_{\max}, s_{\max})^1$.

Proof. We leave this proof out of scope of this paper.

Importantly, cooperative games for either price or quantity competition yield different solutions e.g., product differentiation in the Cournot game.

Algorithm 1. Algorithm based on the Bertrand price competition

---

1: $s_1^* = s_{\max}$
2: $s_2^* = \text{MaximizeProfit2}(s_1^*)$
3: $\mathbf{p}^* = \text{FindOptimalPrices}(s_1, s_2)$
4: **function** MAXIMIZEPROFIT2(s_1^*) ▷ Maximize profit of the SP 2 by s_2
5:     **return** $s_2^* = \arg\max_{s_2} \text{Profit } i(s_1^*, s_2)$
6: **function** PROFIT $i(s_1, s_2)$ ▷ Profit of the SP based on the optimal prices
7:     $\mathbf{p}^* = \text{FindOptimalPrices}(s_1, s_2)$
8:     **return** $\Pi_i(s_1, s_2; \mathbf{p}^*)$
9: **function** FINDOPTIMALPRICES(s_1, s_2) ▷ Prices maximizing the profit for fixed $\mathbf{s}$
10:     **return** $\mathbf{p}^*$: solution of the system (14)

---

5 Numerical Results and Conclusion

In total, we analyze four scenarios: Bertrand and Cournot competition for both the conventional and the realistic distribution each. Even though our approach is suitable for any cost function, for the sake of comparison this section considers the same linear costs for all of the cases. Minding a multitude of possible choices, below we only provide several representative examples for comparison.

We remind that for a particular distribution we quantify the following parameters in our model: the maximum quality $s_{\max}$ (set by default to 100), the cost coefficient v (0.1), and the "richest" customer $\theta_{\max}$ (6.6). In Fig. 2a–c, we illustrate the evolution of our market for varying $s_{\max}$. As it is demonstrated in Fig. 2a, the equilibrium quality for both the uniform (UD) and the exponential (ED) distribution (with $\lambda = 5$) behaves similarly and confirms an identical choice for the Cournot game as well as a clear product differentiation for the Bertrand game. Importantly, we note that the latter results in the same quality for both taste distributions $h(\theta)$.

---

[1] We remind that if $s_1^* = s_2^*$ then $p_1^* = p_2^*$, and the active customers with the positive utility are indifferent to choosing either of the SPs. In this case, the demand is equally shared between the SPs and leads to equal market indicators for them.

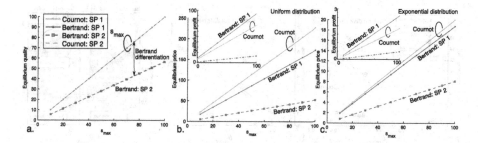

Fig. 2. Evolution of equilibrium indicators for maximum quality s_{max}: (a) equilibrium quality for both distributions, (b, c) equilibrium price and quality for UD and ED.

Further, Fig. 2b, c highlight the difference in prices and profits of the SPs. Intuitively, on a market where the majority of customers are "poor" (ED) the equilibrium prices as well as the profits appear to be much lower. The Cournot competition results in prices that are generally higher than those in the Bertrand competition, but for the ED market this difference diminishes together with the degree of price differentiation between the SPs.

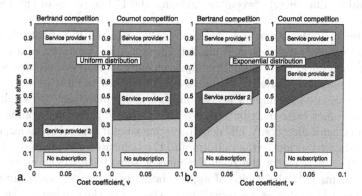

Fig. 3. Evolution of market shares vs. cost coefficient v: Bertrand and Cournot game for (a) UD and (b) ED.

Then, we investigate the impact of costs on the total demand of the SP 1, the SP 2, as well as the share of the market that is not covered. In Fig. 3a, b, we observe the volume of the market that belongs to either of these three groups. While the "wealthier" UD market is less sensitive to changes, on the ED market an increase in costs entails a rise of the equilibrium price as well as a dramatic reduction in the market shares of SPs. Customer churn eventually leads to a significant decrease in the SP profits.

Finally, we analyze all four scenarios in question by varying θ_{max}, which determines the "richest" customer on the market. As for the ED, the market shares stabilize with the growing range of taste, whereas for the UD the market

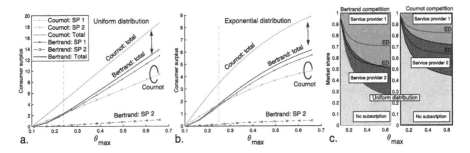

Fig. 4. Market evolution for variable restricting parameter θ_{max}: (a, b) consumer surplus for UD and ED, and (c) market shares.

broadens significantly by covering more and more customers (see Fig. 4c, where dotted lines correspond to the ED market). Further, in Fig. 4a, b for the UD and the ED, respectively, we may observe the total consumer surplus and the separate components for customers of the SP 1 and the SP 2. The relative differences are rather marginal and suggest that the Cournot game is more beneficial for the market than the Bertrand game. However, the absolute values indicate a considerable difference between the UD and the ED in terms of the resultant benefits.

In summary, this paper considered both the price and the quantity competition in a vertically differentiated market. In particular, we analyzed a tractable example with linear costs of quality improvement and proposed a numerical procedure to relax the restrictive assumptions. In contrast to most past studies, we not only evaluated the mobile service market under more realistic assumptions on the customer taste distribution, but also provided a detailed comparison of the key market indicators. While demonstrating similar general behavior, the two considered distributions – the uniform and the truncated exponential – indicate a dramatic difference in the resulting market sensitivity to the changes. The latter confirms that the choice of appropriate customer taste distribution is a crucial factor in analyzing a competitive market, while the general market trends could be understood from simpler assumptions.

References

1. Deloitte Global: Telecommunications industry outlook 2017, US (2017)
2. Liu, Q., Zhang, D.: Dynamic pricing competition with strategic customers under vertical product differentiation. Manag. Sci. **59**(1), 84–101 (2013)
3. Gabszewicz, J.J., Thisse, J.-F.: Price competition, quality and income disparities. J. Econ. Theor. **20**(3), 340–359 (1979)
4. Lancaster, K.: The economics of product variety: a survey. Mark. Sci. **9**(3), 189–206 (1990)
5. Ren, S., Park, J., Van Der Schaar, M.: Entry and spectrum sharing scheme selection in femtocell communications markets. IEEE/ACM Trans. Netw. **21**(1), 218–232 (2013)

6. Niyato, D., Hossain, E.: Dynamics of network selection in heterogeneous wireless networks: an evolutionary game approach. IEEE Trans. Veh. Technol. **58**(4), 2008–2017 (2009)
7. Nguyen, T., Zhou, H., Berry, R.A., Honig, M.L., Vohra, R.: The cost of free spectrum. Oper. Res. **64**(6), 1217–1229 (2016)
8. Bonanno, G.: Vertical differentiation with Cournot competition. Econ. Notes **15**(2), 68–91 (1986)
9. Shaked, A., Sutton, J.: Relaxing price competition through product differentiation. Rev. Econ. Stud. 3–13 (1982)
10. Donnenfeld, S., Weber, S.: Limit qualities and entry deterrence. RAND J. Econ. 113–130 (1995)
11. Choi, C.J., Shin, H.S.: A comment on a model of vertical product differentiation. J. Ind. Econ. 229–231 (1992)
12. Shaked, A., Sutton, J.: Natural oligopolies. Econometrica J. Econ. Soc. 1469–1483 (1983)
13. Motta, M.: Endogenous quality choice: price vs. quantity competition. J. Ind. Econ. 113–131 (1993)
14. Lehmann-Grube, U.: Strategic choice of quality when quality is costly: the persistence of the high-quality advantage. RAND J. Econ. 372–384 (1997)
15. Tirole, J.: The Theory of Industrial Organization. MIT press, Cambridge (1988)
16. Bennis, M., Lara, J., Tolli, A.: Non-cooperative operators in a game-theoretic framework. In: Proceedings of the International Symposium on Personal, Indoor and Mobile Radio Communications. IEEE (2008)
17. Niyato, D., Hossain, E.: Competitive pricing for spectrum sharing in cognitive radio networks: dynamic game, inefficiency of Nash equilibrium, and collusion. IEEE J. Sel. Areas Commun. **26**(1), 192–202 (2008)
18. Gladkova, M.A., Zenkevich, N.A., Sorokina, A.A.: Method of integrated service quality evaluation and choice and its realization on the market of mobile operations of Saint-Petersburg. Vestnik of Saint Petersburg University. Management Series, no. 3 (2011)

Risk Management Using Cyber-Threat Information Sharing and Cyber-Insurance

Deepak K. Tosh[1]([⊠]), Sachin Shetty[2], Shamik Sengupta[3], Jay P. Kesan[4], and Charles A. Kamhoua[5]

[1] Department of Computer Science, Norfolk State University, Norfolk, VA, USA
dktosh@nsu.edu
[2] Virginia Modeling Analysis and Simulation Center,
Old Dominion University, Virginia, VA, USA
sshetty@odu.edu
[3] Department of Computer Science and Engineering,
University of Nevada, Reno, NV, USA
ssengupta@unr.edu
[4] College of Law, University of Illinois, Urbana Champaign, IL, USA
kesan@illinois.edu
[5] Cyber Assurance Branch, Air Force Research Laboratory, Rome, NY, USA
charles.kamhoua.1@us.af.mil

Abstract. Critical infrastructure systems spanning from transportation to nuclear operations are vulnerable to cyber attacks. Cyber-insurance and cyber-threat information sharing are two prominent mechanisms to defend cybersecurity issues proactively. However, standardization and realization of these choices have many bottlenecks. In this paper, we discuss the benefits and importance of cybersecurity information sharing and cyber-insurance in the current cyber-warfare situation. We model a standard game theoretic participation model for cybersecurity information exchange (CYBEX) and discuss the applicability of economic tools in addressing important issues related to CYBEX and cyber-insurance. We also pose several open research challenges, which need to be addressed for developing a robust cyber-risk management capability.

Keywords: Cybersecurity information sharing · Cyber-insurance · Cyber-threat intelligence · Cyber Security Information Sharing Act (CISA)

1 Introduction

Despite the enormous efforts from security researchers, government agencies, and industries toward developing robust security solutions, intelligent adversaries

Approved for Public Release; Distribution Unlimited: 88ABW-2017-2157, Dated: 04 May 2017. This work was supported by Office of the Assistant Secretary of defense for Research and Engineering (OASD (R&E)) agreement FA8750-15-2-0120, Department of Homeland Security Grant 2015-ST-061-CIRC01 and National Science Foundation (NSF) Award #1528167.

© ICST Institute for Computer Sciences, Social Informatics and Telecommunications Engineering 2017
L. Duan et al. (Eds.): GameNets 2017, LNICST 212, pp. 154–164, 2017.
DOI: 10.1007/978-3-319-67540-4_14

find their way in with advanced exploits. Cyber breaches have expanded their breadth not only in the financial sector but also in healthcare, government, educational, defense, and transportation sectors. It was reported that 75% of top 20 financial corporations (banks) are affected by various malwares [1] and some instances include 2014 JP Morgan Data Breach, 2012 DDoS attacks and 2016 SWIFT hack [2]. Losses due to cyber crimes are increasing at an alarming rate and expected to reach $6 trillion by 2021 [3].

In order to abate the impacts of cyber attacks, the organizations, governments, and policy makers are investigating the criticality of ongoing cyber war and proposing mechanisms to effectively defend cyber attacks. The Cybersecurity National Action Plan (CNAP) from U.S. government was proposed in the year 2016 to come up with long-term strategies for fostering cybersecurity awareness, maintain public safety, and protect privacy. The initiative includes establishment of national cybersecurity commission, modernizing government IT infrastructure, and invest more than $19 billion toward cybersecurity research [4]. Besides the efforts from federal level, it must be a customary to adopt best cybersecurity practices at an organizational/individual level. Thus, organizations require the most up-to-date information about attack incidents so as to take proactive measures toward fostering security awareness and better understanding the threat landscape. Since the intelligent attackers can tactfully modify the existing exploits and reuse these exploits for attacking multiple targets, the organizations must collaborate with each other by sharing their vulnerability related information to derive Cyber-Threat Intelligence (CTI) for preventing similar cyber attacks that another firm might have already seen. The bill from U.S congress, "S.754-Cybersecurity Information Sharing Act (CISA) of 2015" [5], encourages DHS to develop a sharing process to facilitate real-time exchange of threat indicators and defensive measures [6] in an automated manner. The bill also provides liability protections to the volunteering parties who share their threat information with other entities or government.

Despite this initiative and advantages of cyber-threat information sharing, organizations are hesitant to take part in such process due to several reasons: (1) lack of trust on the incident exchange process since it may enable competitive advantage to the rivals in the market; (2) possibilities of privacy leak including personal and financial data during the process of sharing that may lead malicious participants to exploit the trust relationship; (3) absence of standardized sharing platform on which organizations can rely upon; (4) insecure feeling of organization to participate in the framework due to the fear of reputation loss; (5) absence of incentivization models to attract corporations toward sharing cybersecurity information; (6) possibility of free-riding, where other organizations take advantages of the shared information without giving anything in return. For availing a globally common format for cyber-information sharing, ITU-T (International Telecommunication Union-Telecommunication) has taken the initiative to adopt a framework called CYBersecurity information EXchange (CYBEX) [7]. However, the framework does not address the fundamental issues, such as trust agreements, governance, or any non-technical aspects, of information sharing.

By addressing these challenges, it can be expected that organizations would be inclined to participate in the threat exchange process so as to strengthen their proactive defense capabilities. At the same time, participation may bring positive externality effect and thereby reducing the investments toward cyber-insurance and self-security expenses.

In this paper, we investigate the need of both cyber-insurance and cyber-security information sharing in developing a resilient cyberspace for the organizations. We provide the motivations and incremental progresses in this area over the recent past years and discuss how economic models are applicable in addressing several crucial problems related to cyber-insurance and CTI sharing. Given the organizations could reap real-time cyber-related knowledge out of the sharing capability, we discuss how an organization's participation decision can be captured using game theoretic approach. Also, we provide a 2-player game model that aims to resolve the trade-off of deciding whether to participate in CYBEX and share or not. We also present several other research challenges that are yet to be addressed.

The paper is organized as follows. We briefly discuss about the background research in Sect. 2. Need of cyber-insurance and cybersecurity information sharing is presented in Sect. 3. Section 4 presents a sample participation game model and some open research challenges are posed in Sect. 5. Finally, Sect. 6 concludes the paper.

2 Related Works

This topic has gained significant attention and is being investigated by government, policy makers, economists, non-profit organizations, industries, cybersecurity and network professionals with researches in this particular area still emerging [8–10]. Considering the need of cybersecurity information sharing, Gordon et al. [11] analyzed the economic (dis)advantages of this activity and derived its relationship with accounting aspects of an organization Through game theoretic model, they prove that such exchange activity improves the social welfare as well as security level of the firms at a reduced expenditure. Furthermore, an incentive mechanism is provided to eliminate the free-rider problem so that no firm can gain more by making under-investment. It is trivial that nature of information plays a major role in deciding economic losses of an organization, however this component was not addressed in [11]. Authors of [12] have proposed a similar game theoretic model to determine the IT security investment levels and compare it with the outcome of a decision theoretic approach that considers various components, such as vulnerability, payoff from investment etc.

Organizations, especially small scale enterprises, are bounded by a limited budget toward cybersecurity, which is why it is necessary to determine the impact of CTI sharing on the investments altogether. Therefore, authors in [13,14] study to determine the optimal expenditure amount in presence of cyber-information exchange that assists organizations in maximizing their overall payoff. Research works presented in [15,16] have looked into this problem by considering a centralized social planner that guides the organizations in choosing the above mentioned decision parameters so as to maximize their social welfare. Departing from the traditional inter-networked cyber users, authors of [17] model a non-cooperative game to analyze decision of security investment and information sharing in cloud computing domain, where virtual machines reside on a common hypervisor and there exists possibility of side-channel attacks.

On the other hand, cyber-insurance market is emerging [18] due to the high occurrence of targeted cyber breaches over the years. However, the components such as interdependent security, correlated risks, and information asymmetries [19,20] make it challenging to model appropriate policies for the organizations. Nash equilibrium analysis and social optima concepts are applied to model security games in [21] that consider above three components into account and decide how investment can be used for both public good (protection) and a private good (insurance). Full insurance and partial insurance coverage models are proposed in [22] and study the impact of cooperation on self-defense investments. Another quantitative framework is proposed in [23] that applies optimization technique to provide suggestions to the network users and operators on investments toward cybersecurity insurance by minimizing the overall cyber risks. Although both of the risk reduction strategies are promising in nature there are several avenues that are untouched and yet to be explored.

3   Information Exchange for Balancing Privacy and Security in Cyber-insurance Market

Cyber-insurance preserves market autonomy and is designed to provide coverages for insureds experiencing losses from cyberspace incidents. The premiums for coverages are determined based on insurance applicants underwriting characteristics, which are the key factors chosen by insurers as indicators of applicants risk levels. The cyber-insurance market is characterized by volatile revenue growth, high demand, low capacity and covered loss is much smaller than total loss. However, the unique issue that many cyber-insurance providers are facing is that information regarding the insured is very opaque to insurers and the link

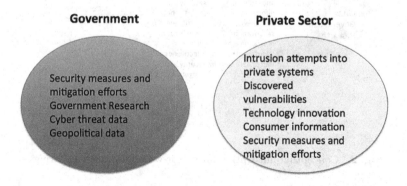

Fig. 1. Public private partnership

between cyber-incidents and financial losses is not well established. There is a lack of quantitative cyber risk assessment and a lot of underwriting is done based on results from questionnaires and interviews [24].

The need to balance privacy and security for facilitating the sharing cyber incidents has generated several debates of legal policy. Private information is held by both the private and public sectors separately and secured to the maximum extent. Any information sharing framework should consider the categories of private information held by both sectors, and the information-sharing program would be narrowly tailored to emphasize the categories of information that would be the most useful to the other side for improving cybersecurity, while excluding the categories of information that would put privacy or national security at risk. Figure 1 illustrates current status of open information sharing [24] and the possible future of open information sharing under a regime like CISA.

Figure 1 illustrates examples of types of information that the different sectors might wish to keep secret [24]. However, in the interest of national security, some types of information would routinely be withheld. For example, while an agency may be forthcoming about recent attempts to hack into its systems, it may be a bad idea to give too much information about the specific vulnerability that was exploited. A privately owned utility company might benefit from information about the vulnerability, but the current paradigm does not have an efficient mechanism for public-private cooperation in cyber-threat information sharing. Information in the right circle could be accessible to the government through existing legal processes. The reluctance to share may be because it could harm a company's reputation or make them into a more attractive target for hackers. This is a major reason why we encourage an organized and largely anonymized system for exchange of vulnerabilities and intrusions.

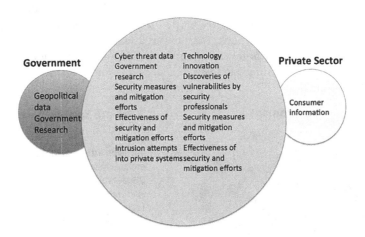

Fig. 2. Conceptual illustration for sector-wide information sharing

Figure 2 illustrates the conceptual illustration for combining the right types of private information without overshare to create a circle of trust [24]. We characterize the middle circle of Fig. 2 as representing a circle of trust managed by a trusted third party. As visualized in Fig. 2, this conceptual illustration would maintain government secrecy for classified military activities and geopolitical information, and would maintain private market secrecy for consumer information, including information about consumers' online activities. In the middle oval, we have placed the types of information that we think could provide the clearest benefits to each sector when shared. Private cybersecurity researchers could benefit from information about intrusion attempts and details about vulnerabilities uncovered by government actors. Government agencies could benefit from up-to-date information about private cybersecurity innovations and the identification of vulnerabilities by private firms. Both sides could benefit from information about different security measures and their rate of success. Some existing laws would need to be revised to implement this proposal, such as the Electronic Communications Privacy Act, which currently may limit the ability of security researchers to share information between firms or with the government.

4   Game Theoretic Model of CYBEX Participation

This Section presents a game model to demonstrate how cybersecurity decisions of interacting organizations are addressable using economic analysis. Despite of understanding the benefits of CTI sharing, most of the organizations are not so motivated to take the risk of participating in CYBEX. Thus, the participation decision requires to be resolved using a cost-benefit approach.

CYBEX Participation Game Model [25]:

In this model, a pair of rational organizations interact with each other to decide whether to participate in the CYBEX or not. Here, CYBEX is a governing entity in the system that imposes participation costs/incentives on the firms to induce participation. The necessity of game theory comes to resolve the following hidden conflict. If CYBEX charges high participation cost, the firms may get deterred from participation, eventually reducing CYBEX's revenue. Whereas, if CYBEX charges too low to attract firms, the revenue generated by CYBEX might be insufficient to sustain in the market. The generic payoff model for organizations must include following two components.

Sharing and Investment Gain: Since organizations are assumed to invest for their own cybersecurity R&D, and infrastructure (firewall, antivirus, and other security products), they receive a direct benefit in term of reduced amount of cyber attacks. Furthermore, the organizations also take advantage of the shared information that leads to additional sharing benefit, which helps to strengthen a firm's proactive cyber-defense capabilities. Subjectively, this benefit comes out of the assistance in strengthening an organization's proactive defense from the received information about vulnerabilities, patches, and fixes.

Cost Components: The involvement in CYBEX requires a participation cost that is imposed by CYBEX to maintain and restrict its utilization by providing liability protections to the firms. In addition to that, sharing of cyber-information has a cost associated which may refer to the combination of extra efforts needed in preparing the information to share and reputation loss incurred due to sharing.

Participation Game in Strategic Form: The participation game can be formalized in a strategic form presented in Table 1, where each firm has the binary strategy set $SS = \{$Participate and Share in CYBEX, Not Participate$\}$.

Table 1. Payoffs in strategic-form for participation game

	Participate & Share	Not Participate
Participate & Share	$Sa\log(1+I) - x - c,$ $Sa\log(1+I) - x - c$	$a\log(1+I) - x - c,$ $a\log(1+I)$
Not Participate	$a\log(1+I),$ $a\log(1+I) - x - c$	$a\log(1+I),$ $a\log(1+I)$

From the Table 1, we can observe that when the interacting organizations are not participating, their benefits come only from the self-investment, which is presented in a logarithmic variant, $a\log(1+I) > 0$, where I is the investment amount and a is a scaling parameter. When both organizations take part in the information exchange, they benefit from sharing as well as self-investment but at a cost of participation (c) and information sharing (x). The combined reward is $Sa\log(1+I)$, where S represents the sharing benefits and assumed to be greater than 1. The top-right and bottom-left corners of the table refers to the payoff scenario when one of the organization does not participate while other one does. Thus, the one who is not participating gets reward only out of its own investment, while the participating firm pays for the participation and sharing but gets no sharing benefits in return.

Analysis: By conducting best response analysis, we can observe that irrespective of what strategy the row player takes, the column player's best strategy depends on the choice of sharing benefits (S) and the cost components. Thus, if cost of participation and sharing dominates the total reward, then organizations will preferably opt for the risk averse strategy of "Not Participate". Then, the Nash Equilibrium (NE) solution of the single-shot game will be ("Not Participate", "Not Participate"). However, the single stage scenario does not apply in practice, rather the organizations take time to figure out the long term optimal strategy. Considering the CYBEX is interested in enabling full participation in the system, incentives are necessary to motivate the players to participate. The detailed analysis of such multistage evolutionary model along with incentivization scheme is given in our prior work [26]. However, we feel that this research needs further extension by relaxing some of the natural constraints assumed in the prior works. In the following, we briefly discuss on various avenues to broaden the scope of this model.

Discussions: The extension ideas are numerated in the following. (E1) In the above model, it is assumed that the organizations have a fixed investment toward security. However, in reality such assumption may not hold true. Therefore, it would be interesting to analyze the participation scenario, when organizations have a differentiated cyber-investments and the amount of information sharing is no longer homogeneous. (E2) The cost of information sharing may not be straightforward as it is depicted in the game, rather a concrete cost model with consideration of attack possibility and privacy would make the case more realistic. (E3) Since some organizations may not be truthful regarding their sharing, this fact will impact the overall participation in the system. Therefore, rigorous analysis is necessary to understand the limits and bounds of maliciousness during information exchange to ensure sustainability of the sharing system.

5   Open Research Challenges

Besides the above directions to extend the CYBEX participation model, there are several challenges exist, which indirectly affect the information sharing decisions of organizations. In the following, we briefly discuss some of these issues.

- **Insurance based mechanism for information sharing:** The participation cost may exhibit the characteristics of insurance which may be a cost or incentive and can be used to motivate socially optimal sharing behavior (through "carrot" incentives like liability protections). However, due to the limited academic literature on cyber-insurance, understanding the effectiveness of cyber-insurance as an incentive/deterrence to induce sharing behavior has become challenging. Also, it is required to know, how long incentives may be applied to develop the sharing attitude without any external incentive.
  To model cyber-insurance, the coverage and premium for the insurance will depend on the sharing level, frequency of cyber attack, and attack severity level. As the frequency of attack increases the premium for the insurance gets incremented compared to previous cycle, however periodically the premium amount decreases depending on how successfully the an organization strives against cyber attacks with the help of cooperation. In the following we present a direction toward premium function $C_{prm}(t)$ which can be used to model the expected premium amount that an organizations need to pay towards insurance.

$$C_{prm}^{t} = \begin{cases} C_{prm}^{t-1} - \delta^{-\alpha_1 t} & \text{if no attack until t and } C_{prm}^{t-1} \geq C_{thres} + \delta^{-\alpha_1 t} \\ C_{prm}^{t-1} + \delta^{\frac{\alpha_2 d}{t_{diff}}} & \text{if attack occured at } t \end{cases}$$

where t_{diff} is the time gap between current time and the last occurrence of cyber attack, δ is the premium exponent defined by the insurance provider, d

is the severity level of the cyber attack and $C_{thres} > 0$, is the min. mandatory premium amount that must be charged to an organization by the insurance agency. $C_{prm}^0 = c_0$ is the initial premium amount decided by mutual understanding of both organization and insurance company.

Two primary challenges in designing such a cyber-insurance mechanism are (i) uncertainty (incomplete information) about the information disclosure and (ii) enforcing truthfulness on information exchange, especially in the case when each organization pays differently based on their reliability and reputations.

- **CYBEX with incomplete information:** What if the firms have only partial or incomplete information in this game? How will the competition evolve if some common information now varies or only an estimate is available to all the players in this game? Thus, it becomes important to also consider these assumptions in while making sharing decisions. While, in the above scenario, we emphasized on fixed investment and "participation" vs. "no participation" with pure strategy, it also becomes necessary to extend the game model to consider possibility of continuous domain of investment ($0 < I_i < I_{max}$) as well as mixed strategy for the firms' participation inclination depending on their feedback from the previous stages and payoffs.

- **Measuring cyber risk:** Cyberinsurance has been recognized as an effective way to improve resilience because it speeds up the process of recovery from financial losses after major cyber attack incidents. It also serves as a complement to self-protection as it creates financial incentives for the insured to mitigate cyber-risks in their systems. The cyberinsurance market is premised on being able to develop a comprehensive understanding and assessment of cyber risk. Lack of measurable cyber risks will hinder the ability to develop policies commensurate with the risk profile.

- **Information asymmetry (Adverse selection):** Companies with poor self protection need insurance to have risks covered. However, it is difficult to distinguish the companies with different self-protection and cyber-risks. There needs to be incentive for companies to share such information. If not, insurer will charge premium based on high risk standard to reduce losses. Thereby, the expensive premium will drive away low-risk companies, which will eventually lead to remaining policies in insurer's portfolio filled with bad risk pooling.

- **Information asymmetry (Moral hazard):** Upon receiving coverage, the policyholder may alter its risk characteristics by reducing self-protection to cut cost. After a loss event, policyholder may ask the insurer to pay unnecessary but covered costs. Hence, there are need for game theoretic approaches to address the moral hazard and adverse selection problems.

6   Concluding Remarks

Traditional management of cybersecurity risks requires a strong taskforce and heavy security investment. However, the traditional approaches are more of reactive in nature. Adopting collaborative approach of cyber-threat information sharing could potentially help the organizations to stay on top of the cyber risks.

Furthermore, cyber-insurance could help in transferring risks to the third-party insurers. While both approaches look promising, there exists several research issues that are unresolved. In addition to discussing the advantages these two risk management methods could bring, we have presented the applicability of game theory in addressing CYBEX participation problem. The open research challenges related to these two mechanisms are briefly discussed to further extend the scope of cybersecurity research and particularly CTI sharing.

References

1. https://cdn2.hubspot.net/hubfs/533449/SecurityScorecard_2016_Financial_Report.pdf
2. https://sentinelone.com/blogs/the-most-devastating-cyber-attacks-on-banks/
3. http://cybersecurityventures.com/hackerpocalypse-cybercrime-report-2016/
4. https://obamawhitehouse.archives.gov/the-press-office/2016/02/09/fact-sheet-cybersecurity-national-action-plan
5. https://www.congress.gov/bill/114th-congress/senate-bill/754
6. Fischer, E., Liu, E., Rollins, J., Theohary, C.: The 2013 cybersecurity executive order: overview and considerations for congress (2013)
7. Rutkowski, A., Kadobayashi, Y., Furey, I., Rajnovic, D., Martin, R., Takahashi, T., Schultz, C., Reid, G., Schudel, G., Hird, M., Adegbite, S.: Cybex: the cybersecurity information exchange framework (x.1500). SIGCOMM Comput. Commun. Rev. **40**, 59–64 (2010)
8. Wang, T., Kannan, K.N., Ulmer, J.R.: The association between the disclosure and the realization of information security risk factors. Inf. Syst. Res. **24**(2), 201–218 (2013)
9. Dandurand, L., Serrano, O.S.: Towards improved cyber security information sharing. In: 5th International Conference on Cyber Conflict, pp. 1–16. IEEE (2013)
10. de Fuentes, J.M., González-Manzano, L., Tapiador, J., Peris-Lopez, P.: Pracis: privacy-preserving and aggregatable cybersecurity information sharing. Comput. Secur. **69**, 127–141 (2016). doi:10.1016/j.cose.2016.12.011. ISSN 0167-4048
11. Gordon, L.A., Loeb, M.P., Lucyshyn, W.: Sharing information on computer systems security: an economic analysis. J. Acc. Publ. Policy **22**(6), 461–485 (2003)
12. Cavusoglu, H., Raghunathan, S., Yue, W.T.: Decision-theoretic and game-theoretic approaches to it security investment. J. Manag. Inf. Syst **25**(2), 281–304 (2008)
13. Tosh, D.K., Sengupta, S., Mukhopadhyay, S., Kamhoua, C., Kwiat, K.: Game theoretic modeling to enforce security information sharing among firms. In: IEEE 2nd International Conference on Cyber Security and Cloud Computing (CSCloud), pp. 7–12 (2015)
14. Tosh, D.k., Molloy, M., Sengupta, S., Kamhoua, C.A., Kwiat, K.A.: Cyber-investment and cyber-information exchange decision modeling. In: IEEE 7th International Symposium on Cyberspace Safety and Security, pp. 1219–1224 (2015)
15. Hausken, K.: A strategic analysis of information sharing among cyber hackers. JISTEM-J. Inf. Syst. Technol. Manag **12**(2), 245–270 (2015)
16. Gal-Or, E., Ghose, A.: The economic consequences of sharing security information. Econ. inf. secur **12**, 95–105 (2004)
17. Kamhoua, C., Martin, A., Tosh, D.K., Kwiat, K., Heitzenrater, C., Sengupta, S.: Cyber-threats information sharing in cloud computing: a game theoretic approach. In: IEEE 2nd International Conference on Cyber Security and Cloud Computing (CSCloud), pp. 382–389 (2015)

18. http://www.businessinsurance.com/article/20161207/NEWS06/912310865/ Cyber-insurance-market-to-grow-says-Allied-Market-Research

19. Anderson, R., Moore, T.: The economics of information security. Science **314**(5799), 610–613 (2006)

20. Böhme, R., Schwartz, G., et al.: Modeling cyber-insurance: towards a unifying framework. In: WEIS(2010)

21. Grossklags, J., Christin, N., Chuang, J.: Secure or insure?: a game-theoretic analysis of information security games. In: Proceedings of the 17th international conference on World Wide Web, pp. 209–218. ACM (2008)

22. Pal, R., Golubchik, L.: Analyzing self-defense investments in internet security under cyber-insurance coverage. In: 2010 IEEE 30th International Conference on Distributed Computing Systems (ICDCS), pp. 339–347. IEEE (2010)

23. Young, D., Lopez, J., Rice, M., Ramsey, B., McTasney, R.: A framework for incorporating insurance in critical infrastructure cyber risk strategies. Int. J. Crit. Infrastruct. Prot. **14**, 43–57 (2016)

24. Kesan, J.P., Hayes, C.M.: Creating a circle of trust to further digital privacy and cybersecurity goals, Mich. St. L. Rev., p. 1475 (2014)

25. Tosh, D.K., Sengupta, S., Kamhoua, C.A., Kwiat, K.A., Martin, A.: An evolutionary game-theoretic framework for cyber-threat information sharing. In: IEEE International Conference on Communications, ICC, pp. 7341–7346 (2015)

26. Tosh, D., Sengupta, S., Kamhoua, C.A., Kwiat, K.A.: Establishing evolutionary game models for cyber security information exchange (CYBEX). J. Comput. Syst. Sci. (19 October 2016). doi:10.1016/j.jcss.2016.08.005. ISSN 0022-0000

Paradoxes in a Multi-criteria Routing Game

Amina Boukoftane[1], Eitan Altman[2,3]([⊠]), Majed Haddad[4], and Nadia Oukid[1]

[1] University of Blida 1, Route de Soumaa, B.P. 270, Blida, Algeria
[2] Universit Cote d'Azur (UCA), INRIA, 06902 Sophia-Antipolis, France
eitan.altman@inria.fr
[3] LINCS, 23 avenue d'Italie, 75013 Paris, France
[4] LIA/CERI, University of Avignon, Agroparc, BP 1228, 84911 Avignon, France

Abstract. In this paper, we consider a routing game in a network that contains lossy links. We consider a multi-objective problem where the players have each a weighted sum of a delay cost and a cost for losses. We compute the equilibrium and optimal solution (which are unique). We discover here in addition to the classical Kameda type paradox another paradoxical behavior in which higher loss rates have a positive impact on delay and therefore higher quality links may cause a worse performance even in the case of a single player.

Keywords: Routing game · Multi-objective problem · Lossy links · Nash equilibrium · Price of anarchy · Paradox

1 Introduction

There has been much work on routing games with additive costs (cost associated with a route is additive over the links of the route) [6,7]. This has been extended to multi-objective additive criteria, see e.g., [4,6,8]. Little is known however on routing games with non additive costs. There has been some work on routing games for some given simple topologies with non-additive costs [6] triggered by networking applications (e.g., [3,5]).

In this work, we consider costs related to weighted sum of two different types of performances: the delay which is additive, and losses which are not. We focus on a simple network model that has been studied in the case of a single objective by [2,3]. We first derive explicit expression of the equilibrium and then study numerically its properties.

In case of a single objective, it has been shown that a Braess type paradox exists for the topology that we consider [2,3]. We identify a new type of paradox which has some surprising behavior. We then compute the price of anarchy (defined as in [1]).

The rest of the paper is organized as follows: In Sect. 2, we describe the system model and the performance measures adopted throughout the paper. In Sect. 3, we compute the global optimum. The Nash equilibrium is computed in Sect. 4. Simulation results along with the discussion are presented in Sect. 6. Section 8 concludes the paper.

© ICST Institute for Computer Sciences, Social Informatics and Telecommunications Engineering 2017
L. Duan et al. (Eds.): GameNets 2017, LNICST 212, pp. 165–172, 2017.
DOI: 10.1007/978-3-319-67540-4_15

2   The Model and Performance Measures

We shall use the load balancing network topology introduced in [2] consisting of three nodes: two source nodes S_r and S_l (r stands for right and l for left) and one common destination node D (see Fig. 1). There are $2N$ sources of flows, $S(i)$, $i = 1, \ldots, 2N$. Each flow consists of an independent Poisson distributed point process with a rate ϕ. Packets from source $i = 1, \ldots, N$ arrive at node S_l (left), whereas packets from source $i = N+1, \ldots, 2N$ arrive at node S_r (right). Source $i = 1, \ldots, N$ can split its flow between its direct path $S_l D$ and the indirect one $S_l S_r D$. Source $i = N+1, \ldots, 2N$ can split its flow between its direct path $S_r D$ and the indirect one $S_r S_l D$.

More precisely, whenever a packet arrives from source i, the source flips a coin that has a probability p_i to have an outcome called "direct" and a probability of $1 - p_i$ to have an outcome called "indirect". If the outcome is "direct" then the packet is routed over the direct route, and otherwise it is routed over the indirect one. The process of packets originating from source i that take the direct path is thus Poisson with rate ϕp_i. The process of packets that arrive at node i and that take the indirect path is Poisson with rate $\phi(1 - p_i)$. Let x_l^i be the rate of flow sent by source i through link l. Links $S_r S_l$ and $S_l S_r$ are assumed to be wireless so that packets sent over $S_r S_l$ and $S_l S_r$ suffer independent losses with some fix probability q. The delay over these links is assumed to be a constant denoted by δ. Links $S_l D$ and $S_r D$ are assumed to be lossless but they incur a congestion cost per flow unit that uses them of $T_l(x) = 1/(C - x_l)$. Here, C is the link capacity and x_l is the total flow going through link l. Therefore for each player i, consider the arrival process of packets that arrive at node S_k and that are rerouted to the indirect path $S_k S_m D$ (where $k = l, r, m \neq k$). It is a Poisson process as well and its rate is $\phi p(1 - p_i)(1 - q)$.

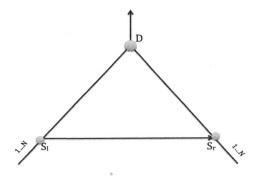

Fig. 1. The competitive routing model.

Let $i \leq N$. The cost for source i is a weighted sum of the average delay of its flow and its loss rate:

$$J_i(p) = \frac{\phi p_i}{C - \phi \sum_{j=1}^{N} p_j - \phi q \sum_{i=N+1}^{2N} (1 - p_j)}$$

$$+ \frac{\phi (1 - q)(1 - p_i)}{C - \phi \sum_{j=N+1}^{2N} p_j - \phi q \sum_{i=1}^{N} (1 - p_j)} \tag{1}$$

$$+ \phi \delta (1 - q)(1 - p_i) + \gamma \phi q (1 - p_i).$$

The three first terms correspond to the delay cost and the last term corresponds to the cost of losses. The first term corresponds to the congestion cost in the direct path of i and the two following terms correspond to the congestion cost along the indirect path. The optimal symmetric solution is obtained by minimizing $\sum_i J_i(p)$ over p_i and adding the constraint that p_i are the same for all i (we then omit i from the notation p_i).

Let

$$X_l(x) = \phi \sum_{j=1}^{N} p_j + \phi q \sum_{j=N+1}^{2N} (1 - p_j)$$

be the rate of packets that use link $S_l D$. We assume that the link cost per unit of flow are linear in the flow through them. This amounts to use the first term in the Taylor's expansion of (1). $J_i(p)$ is thus approximated by

$$J_i(p) = \phi p_i(ax_l + b) + \phi(1 - q)(1 - p_i)(ax_l + b) + \delta \phi(1 - p_i) + \gamma \phi q(1 - p_i)$$

where a and b are some positive constants.

3   Global Optimum Calculation

The global optimal solution is obtained by solving

$$\frac{\partial}{\partial p} \sum_{i=1}^{2N} J_i(p) = 0$$

(unless it is on the boundary). We obtain the unique solution:

$$p = \frac{1}{2} \frac{a\phi N(2q^2 - 2q + 1) + bq - \gamma q - \delta}{\phi a q N(q - 1)}.$$

We find it convenient in the numerical investigation to write $z = \gamma q + \delta$ since the dependence of the global optimum or the equilibrium on each one of the two parameters δ and γ (for fixed q) appears only through the value of z. Thus, p can be written:

$$p = \frac{1}{2} \frac{a\phi N(2q^2 - 2q + 1) + bq - z}{\phi a q N(q - 1)}.$$

Note that

$$J'(p) = 2a\phi^2 q^2 N(1-p) + \phi bq + \phi^2 aN + 2a\phi^2 qN(p-1) - \phi z.$$

$J'(p) = 0$ for

$$z = -2a\phi qN(((1-p)(1-q)) + bq + a\phi N$$

$J'(p)$ is positive if $z < \phi aN(2q(q-1)(1-p)+1) + qb$. In this case J is an increasing function, and the minimum is reached on 0.

On the other hand, if $z > \phi aN(2q(q-1)(1-p)+1) + qb$, J is a decreasing function, and the minimum is reached on 1.

Note that

- $p < 1$ if $z < a\phi N + qb$
- $p > 0$ if $z > a\phi N(2q^2 - 2q + 1) + qb$.

4   Equilibrium Calculation

The equilibrium is obtained by setting $p_i = p$ to be the same for all i except for $i = 1$ where it is taken to be equal to $\hat{p}$. We then find for each value p the best response $\hat{p} = f(p)$ for player 1. A fixed point of this equation provides the equilibrium. We did the same as below and get the Nash equilibrium which is equal to:

$$p_1 = \frac{1}{2} \frac{-a\phi qN(p+q-pq) + q(\phi ap + \phi a - b + \gamma) - \phi a + \delta}{\phi aq}.$$

Let $\hat{p}$ be the point obtained by replacing p_1 by p, we obtain:

$$\hat{p} = \frac{a\phi Nq^2 + a\phi(1-q) + qb - \gamma q - \delta}{a\phi q(qN - N - 1)}.$$

$$\hat{p} = \frac{a\phi Nq^2 + a\phi(1-q) + qb - z}{a\phi q(qN - N - 1)}.$$

We have $qN < N < N + 1$, so the denominator is negative.

By differentiating the cost function with respect to $\hat{p}$ and setting the derivative equal to zero, we get for equilibrium these conditions:

- J is an increasing function if: $z > 2a\phi qN - a\phi N + bq$,
- J is a decreasing function if: $z < 2a\phi qN - a\phi N + bq$.

Also note that

- $\hat{p} < 1$ if $z < a\phi(qN + 1) + bq$
- $\hat{p} > 0$ if $z > a\phi q^2 N + a\phi(1-q) + bq$

5   Price of Anarchy and Paradoxes

The price of anarchy is the ratio of the worst case objective function value of a Nash equilibrium and that of an optimal outcome. That measures how the efficiency of a system degrades due to selfish behavior of its agents. The price of anarchy [1] is a method to measure the inefficiency of equilibrium, it has been used to measure the inefficiency in congestion networks. In this case, each user of the network has a source and destination and they must pay a cost to travel from their source to their destination. In this case it is given by:

$$PoA = \frac{2NJ_i(\hat{p})}{2NJ_i(p)}$$

We say that a paradox occurs if when replacing links with higher quality ones result in worse performance. In our case, a higher quality link could mean a link with smaller delay δ or one with a smaller loss probability q. For example, there is a paradox if the derivative $J'(\delta)$ of J at the equilibrium w.r.t. the delay δ is negative, where

$$J'(\delta) = \frac{a\phi + a\phi N^2(1 + q^2) + 2a\phi Nq(1 - N) + 2bq - 2\delta - 2\gamma q}{aq(qN - N - 1)^2}$$

In a similar network with a single objective for each player, a paradoxes been observed [2,3,5] in which, for suitable parameters, improving the quality of the link(s) between S_r and S_l results in worse performance for all players. We may search for a similar paradox in our problem in which the quality of the link stands for its delay (higher quality means lower delay) or loss rate (higher quality means lower loss rate). The condition for this type of paradox is then that J at equilibrium would be decreasing in the network parameter (e.g., in the delay δ). Thus, the derivative of J at equilibrium should be decreasing where the latter is given by

$$J'(\delta) = \frac{a\phi + a\phi N^2(1 + q^2) + 2a\phi Nq(1 - N) + 2bq - 2z}{aq(qN - N - 1)^2}$$

Note that J is a decreasing function if

$$z > \frac{1}{2}\phi a(1 + N^2) - \frac{1}{2}\phi aqN(2N - qN - 2) + bq$$

6   Numerical Results

Let us now validate our theoretical findings through numerical simulations. We consider $a = 1, b = 1, N = 4, \phi = 1, q = 0.5$. Notice that the domain of existence of z in this case is $[2.2, 3]$.

Figure 2 depicts the equilibrium $\hat{p}$ and the optimal solution p as function of z. As expected, we observe that both $\hat{p}$ and p are increasing in z, while the equilibrium $\hat{p}$ dominates the optimal solution p. Figure 3 depicts the cost function of the optimal solution $J(p)$ and the equilibrium $J(\hat{p})$ as a function of z. We observe that both curves are increasing functions. In Fig. 4, we present respectively the variation of the cost function at the optimal solution and at the equilibrium as a function of the loss probability q. Figure 5 depicts the variation of the cost function at the equilibrium as a function of the link delay δ. The price of anarchy is presented in Fig. 6. As expected, for low values of z, the price of anarchy tends to 1.

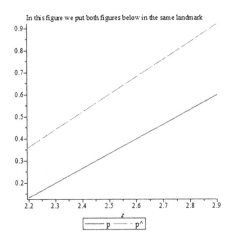

Fig. 2. The optimal solution and the equilibrium as a function of z.

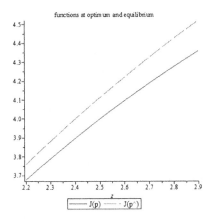

Fig. 3. The cost function at the optimal solution and the equilibrium as a function of z.

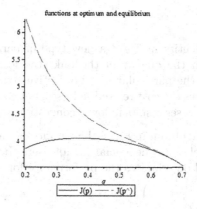

Fig. 4. The cost function J as a function of the loss probability q.

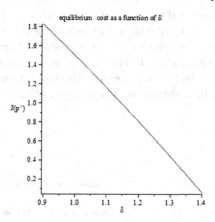

Fig. 5. $J(\hat{p})$ as a function of the link delay δ.

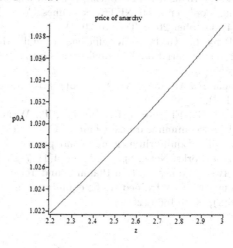

Fig. 6. The price of anarchy.

7 Discussion

New paradox: We identify in Fig. 4 a new type of paradox: the cost is seen not to be monotone in the quality of the link (the loss probability q). This phenomenon is due to the particular multi-objective structure of our problem. Indeed, higher q increases the cost related to losses, but contributes to decreasing the global cost as more losses results in lower congestion and thus in lower delays.

Kameda-paradox: We obtain in Fig. 5 the paradox already observed in [2,3,5] in which larger link delay are beneficial for all users. Investing in faster links increases the delay and deteriorates the performance for all players.

8 Conclusion

In this paper, we have studied the routing game with lossy links and congestion. The cost included both a delay component as well as one corresponding to the losses. After computing the unique optimal solution and the symmetric equilibrium, we have showed that even in the case of global optimization there may be a paradox due to the fact that increasing the loss rate may be advantageous when delays are high. In addition the Kameda-paradox has been also shown to occur here.

References

1. Koutsoupias, E., Papadimitriou, C.: Worst-case equilibria. In: Proceedings of STACS (1999)
2. Kameda, H., Altman, E., Kozawa, T., Hosokawa, Y.: Braess-like paradoxes in distributed computer systems. IEEE Trans. Autom. Control **45**(9), 1687–1690 (2000)
3. Altman, E., Kuri, J., El-Azouzi, R.: A routing game in networks with lossy links. In: 7th International Conference on NETwork Games Control and OPtimization (NETGCOOP 2014), October 2014, Trento (2014)
4. Wu, Y., Peng, Y., Peng, L., Xu, L.: Super efficiency of multicriterion network equilibrium model and vector variational inequality. J. Optim. Theor. Appl. **153**(2), 485–496 (2012)
5. Altman, E., El-Azouzi, R., Abramov, V.: Non-cooperative routing in loss networks. Perform. Eval. **49**(1–4), 43–55 (2002)
6. Altman, E., Boulogne, T., El Azouzi, R., Jimenez, T., Wynter, L.: A survey on networking games in telecommunications. Comput. Oper. Res. **33**, 286–311 (2006)
7. Wynter, L., Altman, E.: Equilibrium, games, and pricing in transportation and telecommunications networks. Netw. Spacial Econ. **4**(1), 7–21 (2004). Special Issue of on Crossovers between Transportation Planning and Telecommunications
8. El-Azouzi, R., Altman, E.: Constrained traffic equilibrium in routing. IEEE Trans. Autom. Control **48**(9), 1656–1660 (2003)

Author Index

Printed in the United States
By Bookmasters